# hong kong success story

As the new Hong Kong International Airport welcomes the world's travellers, we're proud to have played a major part in the whole story — right from leading the far-reaching PADS study which set the framework for Hong Kong's development to 2011 with an initial US$20 billion of new transport infrastructure.

As well as designing the airport terminal itself — with Foster and Partners and BAA plc — we helped bring four more of the principal Airport Core Projects to fruition: Lantau Link with its award-winning Tsing Ma Bridge, North Lantau Expressway, West Kowloon reclamation and Tung Chung new town.

All told, we've helped conceive, plan, design and manage over US$6.5 billion worth of bridges, roads, railways, buildings, utilities and reclaimed land — a resounding success story.

For more details contact
Tony Powlesland at:

Mott MacDonald
St Anne House
Wellesley Road
Croydon CR9 2UL
United Kingdom

T  +44 (0)181 774 2000
F  +44 (0)181 681 5706
E  marketing@mottmac.com
W  www.mottmac.com

## mm Mott MacDonald

# WS Atkins - Comprehensive consultancy services from planning to commissioning and beyond

## The New Hong Kong International Airport at Chek Lap Kok

### WS Atkins' involvement:

**1993** - Awarded the Design Commission, in joint venture, for the southern runway, taxiways and aprons, including master plan review, scheme and detailed design, coordination of all services and drainage.

**1994** - Developed new simulation model to predict settlement performance of multi layered soils beneath reclaimed platform for airport pavements.

**1996** - Awarded the Design Commission, in joint venture, for the northern runway, taxiways and aprons, including all drainage works for the northern area.

**1997 -** Integration of advanced control and communication systems used throughout the Terminal Building and around the airport, providing full system monitoring capability.

**1998** - We are continuing our involvement at Chek Lap Kok with Dynatest International carrying out strength measurements to predict future pavement life. It is planned to incorporate this data in our proposed 'AIRPORTS' Pavement Management System.

# PROCEEDINGS OF THE INSTITUTION OF CIVIL ENGINEERS

# *Hong Kong International Airport, Part 1: airport*

SUPPLEMENT TO CIVIL ENGINEERING
VOL. 126 • SPECIAL ISSUE 1 • 1998

## CONTENTS

*Front cover: The new international
airport at Chek Lap Kok*

# Introduction

*Ernest Irwin*, MBE, BA, BAI, MSc, CEng, FICE

Welcome to this special issue of Civil Engineering, the general journal of Proceedings of the Institution of Civil Engineers. It is the first of two special issues which are intended to provide an overview of the planning, design and construction of Hong Kong's new international airport at Chek Lap Kok and its associated road and rail links. The £12 billion project was completed in July this year on time and within budget and is a testament to Sino-British co-operation.

This issue discusses the planning, management and procurement of the 10 main elements of the so-called Airport Core Programme and then focuses in on the civil engineering design and construction of the 1248 ha airport. Design and construction of the 34 km road and rail link—including the stunning Tsing Ma bridge and Western Harbour Tunnel—is covered in the second special issue scheduled for publication in November 1998.

The first paper in this issue is by Donald Tsang, financial secretary of what is now the Government of the Hong Kong Special Administrative Region of the People's Republic of China. He introduces the main parties involved in the airport development, summarizes the funding arrangements and explains the political context in which the project was undertaken—including the hand over of the region's sovereignty from Britain to China in July 1997.

Billy Lam, director of the New Airport Projects Co-ordination Office, then takes us on an illustrated tour of the ten main elements of the Airport Core Programme, ranging from the airport at Chek Lap Kok island to the dramatic series of bridges, viaducts and tunnels which link it to the Central business district. He also explains the management and control structures, tendering procedures, conditions of contract and dispute resolution mechanisms which were used.

The next paper by Graham Plant and Douglas Oakervee of the Hong Kong Airport Authority describes the civil engineering design for the airport including reclamation works, runways and taxiways, the 550 000 m² passenger terminal building, ground transportation centre and essential supporting infrastructure. Factors which led to the location, size and layout of the airport are discussed as well as the wide range of environmental initiatives which were undertaken.

The final paper is by Alistair Thompson, also from the Airport Authority, and Douglas Oakervee, and describes what happened during the seven-and-a-half year construction programme. It focuses in particular on the unique construction support facilities and services which were provided to support the 21 000 people working on the remote island site during the peak construction phase.

My thanks go to all the authors for their time and patience in preparing and revising their papers, to the referees for reviewing them and in particular to the Hon. Sang Kwong, Secretary for Works, and his colleague Daniel Lam, for co-ordinating the production and delivery of all papers on the ICE's behalf.

*Ernest Irwin, consultant and formerly chairman of civil engineering at Ove Arup Partnership, is Honorary Editor of the ICE Civil Engineering Journal and was responsible for the assessment of all papers in this special issue.*

# Introduction to the Hong Kong Airport Core Programme

*The Hon. Donald Tsang, JP*

The US$ 20 billion Hong Kong Airport Core Programme is one of the largest infrastructure developments in the world. In addition to a brand new airport on reclaimed land capable of handling up to 87 million passengers and 9 Mt of cargo a year, it includes a highly complex 34 km road and rail link to the Central district of Hong Kong and a variety of associated development schemes. Completed on time and to budget despite a background of economic and political upheaval, the project serves as a role model for major public sector infrastructure projects and is a testament to Sino-British co-operation.

*Proc. Instn Civ. Engrs, Civ. Engng, Hong Kong International Airport, Part 1: airport*, 1998, **126**, 3-4

*Paper 11520*

*Written discussion closes 15 November 1998*

Hong Kong achieved a remarkable record of economic success in the past decades. In 1996, its per capita gross domestic product (GDP) reached US$ 24 500 per year, which was ahead of Canada, the UK and Australia. Hong Kong was also the world's seventh largest trading economy, although it ranked only 95th (as at the end of 1994) in terms of population. With few natural resources of its own, Hong Kong thrives as an externally-oriented economy which depends greatly on a modern and ever-expanding infrastructure.

*Fig. 1. Hong Kong's existing Kai Tak airport in Kowloon*

## The need for a new airport

Hong Kong's existing Kai Tak airport, the busiest in the world in terms of international cargo and the third busiest in terms of international passenger throughput in 1996, has been operating at full capacity for some years. Its location in the urban area in a confined site makes further expansion of services there impracticable (Fig. 1).

With a view to satisfying Hong Kong's air traffic needs well into the next century, the Hong Kong government launched the projects in the Airport Core Programme (ACP) in 1990 to provide for a new world-class airport and a fully fledged supporting infrastructure. The new airport, at full development, will be able to handle 87 million

*Donald Tsang is Financial Secretary, Hong Kong*

passengers and 9 Mt of cargo a year; that is, three and six times, respectively, the capacity of the existing airport at Kai Tak (Fig. 2).

### The Airport Core Programme

The ACP, at a cost of HK$ 155·3 billion (US$ 19·9 billion) in money-of-the-day (i.e. out-turn price) terms, ranks as one of the largest infrastructure developments in the world.

The ten projects that make up the programme are providing Hong Kong with a modern airport, a new railway, a new tunnel across the harbour, a new town, new roads and bridges including the magnificent Tsing Ma bridge. They are also creating valuable and new land in the Kowloon peninsula and in Hong Kong's busy Central business district.

### Key parties involved

The Hong Kong government masterminded the implementation of the ACP, working with the Airport Authority (AA) and the Mass Transit Railway Corporation (MTRC), both statutory bodies wholly owned by the government, and a franchisee for the cross-harbour tunnel.

The government reclaims land, builds the transport corridor and creates a new town near the new airport. The AA plans, develops and finally operates the new airport and the MTRC constructs and runs a 34 km long railway between urban Hong Kong and the new airport. Another private sector consortium provides an additional cross-harbour tunnel under a 30-year build-operate-transfer (BOT) franchise.

Overseeing the ten ACP projects and co-ordinating the efforts of the various parties at central government level is the Airport Development Steering Committee (ADSCOM) which sets policy and monitors progress with a government office, the 'New Airport Projects Co-ordination Office' (NAPCO), serving as its executive arm.

### Funding and implementation

The magnitude and complexity of the ACP are unprecedented in Hong Kong, arousing considerable private sector participation and investment. Of the estimated expenditure of HK$ 155·3 billion, the government has provided HK$ 109·9 billion, about 70% of the total cost, by direct funding or through equity injection into the AA and MTRC. The remaining 30% comes from private sector participation in various franchises and commercial loans. The ACP has also spurred extensive real estate developments such as an airport hotel and residential and commercial blocks above airport railway stations.

The constraints of Hong Kong's busy road surface, steeply sloping terrain and the very tight time frame presents daunting engineering challenges when pursuing a massive infrastructure programme such as the ACP. The opening, ahead of schedule and well within budget, of the ACP transport corridor from the business district through expressways and bridges to the Tung Chung new town, some 34

km away, is solid evidence of Hong Kong's ability to manage large projects and to meet stringent targets.

In fact, the ACP budget through the years has been reduced in three stages—in January 1994, November 1996 and November 1997—from the original estimate of HK$ 163·7 billion to the present figure of HK$ 155·3 billion. Key factors underpinning this accomplishment have been rigorous cost control and the open and non-discriminatory tender system, which ensures value for money and encourages international participation. A 'level playing field' is the cornerstone of Hong Kong's economic success and is a principle which has been held in the past and will continue to be maintained.

### The political context

The ACP projects were taken forward in a unique political situation when the change of sovereignty over Hong Kong was on the horizon. A high degree of understanding and co-operation was required between the British and the Chinese governments, especially over issues of debt financing and franchise awards which straddled mid 1997.

Through the signing of a memorandum of understanding (MOU) between the two governments in 1991 and subsequent discussions at the airport committee of the joint liaison group (JLG) established under the 1984 Sino-British joint declaration, various agreements were reached on many important issues critical to ACP progress.

### A success story

Work has been progressing well on the ACP. Eight of the ten projects have been completed ahead of time and well within budget. This would not have been possible without the hard work of those concerned, solid community support and a competitive procurement policy.

The government looks forward with confidence to successful completion of all ACP projects and opening of the airport and airport railway in 1998, which will add to Hong Kong's credentials as a service centre for infrastructure development in the region.

*Fig. 2. The new Hong Kong International airport under construction at Chek Lap Kok*

# Management and procurement of the Hong Kong Airport Core Programme

B. C. L. Lam, OBE, JP, MSc, MCIPS

Proc. Instn Civ. Engrs,
Civ. Engng, Hong Kong
International Airport,
Part 1: airport, 1998,
**126**, 5-14

Paper 11521

*Written discussion closes
15 November 1998*

This paper describes the ten main elements of the US$ 20 billion Hong Kong Airport Core Programme, ranging from the 1248 ha airport platform at Chek Lap Kok to the dramatic series of bridges, viaducts and tunnels which make up the 34 km road and rail link to the Central business district. It discusses the formidable problems and constraints which faced the 35 000 construction workers and explains the complex management and control structures which enabled this huge project to be completed on time and within budget. Tendering procedures, conditions of contract and dispute resolution mechanisms are also reviewed.

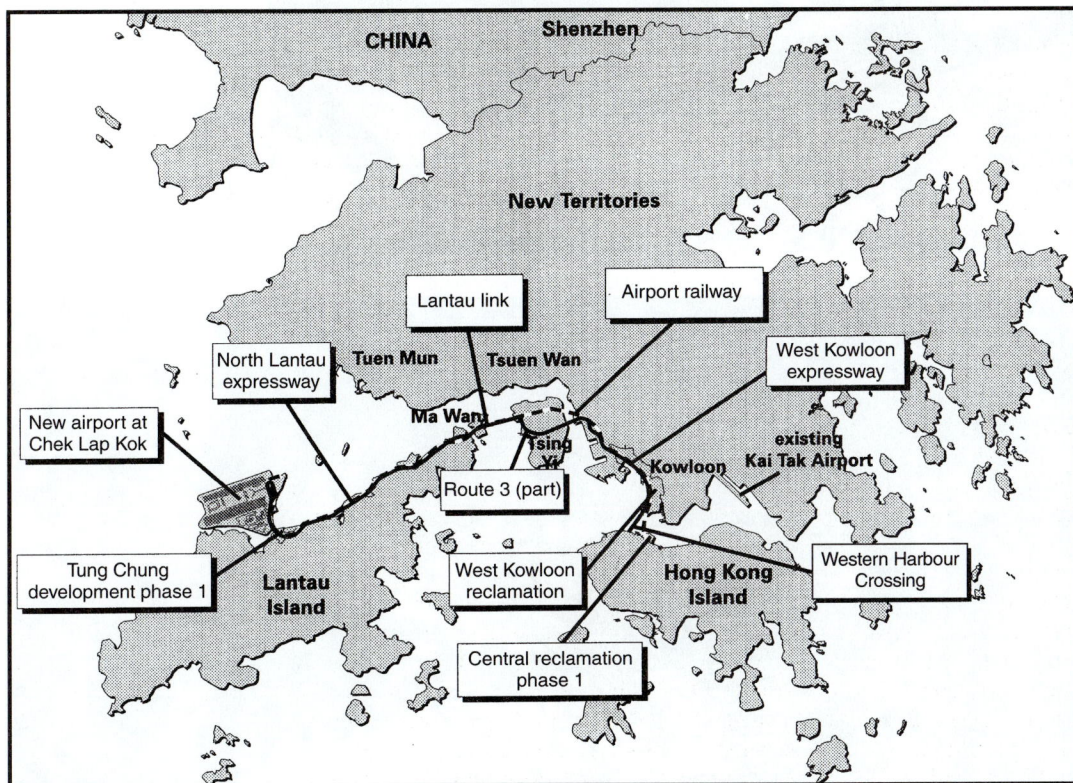

*Fig. 1. Plan of Airport Core Programme elements*

The Hong Kong Airport Core Programme has a number of interesting aspects, as follows.

- In engineering terms, its size (US$ 19·9 billion), diversity of engineering components, very tight time frame and high level of inter-contract dependencies.
- It is of immense political and economic importance to Hong Kong.
- At times it became a point of tension between Britain and China as some of the commitments went beyond the handover of sovereignty in July 1997.

These are the primary factors which shaped the way in which the project has been managed.

## Elements of the project

As anyone who is familiar with Hong Kong is aware, the steeply sloping rocky hills, the fragrant harbour and its adjacent waters and dense urban populations present marvellous physical challenges to civil engineers when it comes to building a new airport and its associated infra-

*Billy Lam was director
of the new airport pro-
jects co-ordination
office in Hong Kong*

*Fig. 2. New airport terminal at Chek Lap Kok*

*Fig. 3. Tung Chung new town adjacent the new airport*

structure. Comprehensive screening of potential sites led to the final selection of Chek Lap Kok, a small island off the north coast of Lantau Island, some 30 km to the west of the Central business district of Hong Kong, as the location for the new airport (Fig. 1). As a consequence, a wide variety of engineering solutions were incorporated into the project, contributing to its interest and drawing together international consultants and contractors with the necessary range of expertise and experience.

The specifically established Airport Authority (AA) and the existing Mass Transit Railway Corporation (MTRC), both wholly owned government bodies, were appointed to develop and operate the new airport and airport railway, respectively, while a private franchise was awarded for the Western Harbour Crossing road tunnel. The remainder of the infrastructure was implemented directly by the Hong Kong government through its public works engineering departments, generally acting as the employer for a series of design consultancies and construction contracts. The private sector was also called upon to co-operate with relocation, particularly relating to the shipyards, dockyards, pumphouses and private piers associated with the urban reclamations at West Kowloon and Hong Kong Central.

The magnitude and complexity of the Airport Core Programme (ACP) was unprecedented in Hong Kong. It comprised ten interrelated projects (over 200 works contracts) being carried out by four separate sponsors—the Hong Kong government, AA, MTRC and the Western Harbour Tunnel Company. Each of the ACP projects represented a large and complex development, which was managed individually and also collectively to meet the prime objective of completion on time and in the most cost-effective manner. The situation was further complicated by the close interface relationships between the projects and the need properly to phase works between them as they progressed in parallel or in sequence.

### The new international airport at Chek Lap Kok

The airport is sited on a 1248 ha platform created by flattening the island of Chek Lap Kok with major reclamations to its west (Fig. 2). The entire works have been progressed directly by the AA with the exception of certain 'stand alone' facilities provided directly by the government such as the air traffic control tower and control centre, and by key franchisees, such as for air cargo, aircraft catering, aviation fuel supply and aircraft maintenance facilities.

### Tung Chung new town development

Initially this was a 60 ha reclamation with supporting roads and utilities implemented by the Territory Development Department (TDD). Public housing construction was undertaken by the Housing Department. This, together with private

*Fig. 4. North Lantau expressway*

property development around Tung Chung station, cater for an initial population of 20 000 (Fig. 3).

### North Lantau expressway

This is a 12·5 km dual three-lane expressway along the shore line of north Lantau Island (Fig. 4), implemented by the Highways Department (HyD). The corridor for the airport railway and a utilities corridor were created in parallel with the expressway and a 30 ha rail depot reclamation was also constructed on behalf of MTRC.

### Lantau link

These are the bridge structures which connect the island of Lantau with Tsing Yi by way of Ma Wan (Fig. 5). Implemented by HyD, the link comprises the 430 m main span Kap Shui Mun cable-stayed bridge, the connecting viaduct across Ma Wan island and the 1377 m main span Tsing Ma suspension bridge. All structures carry six lanes of expressway on the upper deck, with two rail tracks and two emergency vehicle lanes on the lower deck.

*Fig. 5 Lantau link, comprising Kap Shui Mun bridge (foreground), Ma Wan viaduct and Tsing Ma bridge*

### Route 3 (Tsing Yi and Kwai Chung sections)

This is a dual three-lane (twin bore) rock tunnel 1·6 km long taking the expressway from Lantau link through Tsing Yi Island, followed by a 600 m five-span balanced cantilever bridge crossing the Rambler Channel and a 3-km-long dual four-lane elevated expressway skirting the container terminals, implemented by HyD (Fig. 6).

### West Kowloon reclamation

This reclamation, of 334 ha, lies along the western shore line of the Kowloon peninsula, increasing its size by a third; implemented by the TDD and the Civil Engineering Department (Fig. 7). This reclamation was needed to support the airport-related transport corridor and provide land for future development.

### West Kowloon expressway

This is a 4 km dual three-lane expressway implemented by HyD connecting the Kwai Chung viaduct with the Western Harbour Crossing (Fig. 8).

*Fig. 6. Kwai Chung
viaduct*

The northern section is elevated, providing the rail formation at grade beneath, while the southern section is generally at grade, with the airport railway cut-and-cover tunnels incorporated beneath its main interchange.

*Western Harbour Crossing*

This is a 2 km dual three-lane immersed tube crossing the harbour implemented by the Western Harbour Tunnel Co, with the award of a 30 year BOT franchise (Fig. 9).

*Central reclamation*

This 20 ha reclamation on the north shore of Hong Kong Island adjacent to the Central business district, supports the new airport railway Hong Kong station (Fig. 10). It was implemented by TDD, with construction management by MTRC for optimum integration of station construction.

*Airport railway*

This 34 km, 2/4 track, dual-purpose airport express/domestic Lantau line has been implemented by MTRC. Some 70% of the wayleave was provided through HyD contracts owing to the proximity of alignments (Fig. 11). Major stations and associated developments are located at Central, Kowloon, Tai Kok Tsui, Lai King, Tsing Yi, Tung Chung and at the new airport.

## Problems and constraints

As is evident, many of the contributing projects were heavily interrelated, interdependent and were often under the authority of different 'employer' organizations or departments.

The overall available time to the stated completion date for the project was short by any standards, with 95% of the main civil works required to be constructed between mid-1992 and the end of 1996. This necessitated a significant degree of contract overlap and phased handovers to subsequent contracts. At the West Kowloon reclamation, for example, existing waterfront facilities had to remain in operation while new facilities were being built. The new reclamation was progressed in stages, in sufficient time for the expressway and rail builders to phase their works along their common horizontal alignment on this new land.

In some areas, up to three dredging and reclamation contracts and two expressway contracts were working simultaneously. While on an individual basis, environmental compliance had to be ensured, adequate processes had to be in place to pre-empt any 'stop-the-job' situations arising from the aggregate effect of nuisance from multiple sources.

Land for the projects needed to be acquired very quickly through a number of legal routes, depending on its ownership status and, similarly, legal clearance to utilize the sea-bed, either for the dredging of marine sand for reclamation, or for reclaiming upon, had to be processed and established in accordance with statutory provisions.

All structures likely to affect the safety of the public which were not being developed directly by a government works department were subject to buildings department checks and approvals at key stages of design and construction. In a programme such as this, the workload was intense, with delays in approvals being reflected into works progress.

Funding, particularly for the directly financed contracts, was secured from the Finance Committee of the Legislative Council in accordance with a tight legislative timetable, and in accordance with established procedures.

All projects which involved franchises or repayment of borrowings spanning mid 1997 required Chinese endorsement via the airport committee of the joint liaison group with membership from British and Chinese governments.

At the peak of construction activities, some 35 000 workers were employed of which about 5000 were imported workers. Imported labour

*Fig. 7. West Kowloon reclamation*

was only permitted when local shortages could be positively demonstrated in specific trades.

Many construction activities such as site formation, dredging and major bridges carried considerable risk. A comprehensive set of arrangements regarding management structure and policies were in place to promote safety performance to reduce injuries and fatalities.

## Management objectives

The fundamental management objectives were

- to ensure that the necessary actions and responses from various government departments and authorities (Provisional Legislative Council, Finance Bureau, Lands Department, Environmental Protection Department, Buildings Department, etc.) were being delivered in a timely and responsive manner to support the programme fully;
- to establish and maintain certainty of programme and progress, budget and expenditure, together with proactive inter-contract interface management;
- to ensure that political uncertainty, and the project disruptions which can flow from it, were kept to a minimum;
- to ensure open and fair competition for contracts throughout.

## Management structure

In view of the complexity and multi-project nature of the ACP, a high-level management structure (shown in Fig. 12) was set up to ensure its smooth implementation. The main features of this management structure were

- The Airport Development Steering Committee (ADSCOM), with the chief secretary for the Administration as chairman and policy secretaries as members, which had the overall responsibility for overseeing the smooth implementation of the ACP and for co-ordinating actions taken by government with regard to

the memorandum of understanding (MOU).
Significant policy issues and matters affecting
more than one policy bureau were subject to
collective decisions by ADSCOM to ensure a
quick result.

- The individual policy secretaries were responsible for matters concerning the ACP that fell
within their respective policy areas.
- The Finance Bureau was responsible for budgetary control and for ensuring that overall
financing of the ACP satisfied the requirements of the MOU.
- The Subcommittee on Airport Core
Programme (SCACP), chaired by the secretary for works, served to monitor and resolve
all technical works-related problems that
could not be resolved at lower levels.
Problems unresolved by SCACP were
referred to ADSCOM for decision.
- The new Airport Projects Coordination Office
(NAPCO), as the executive arm of ADSCOM,
was responsible for the overall management of
project implementation and co-ordination, and
the government's public information and community involvement programmes, including
servicing the Airport Consultative Committee.
NAPCO had two specific roles to play in project management, programme management
and provision of general support services. On
programme management, NAPCO co-ordinat-
ed and provided guidance to implementing
departments and agencies as required to
ensure early resolution of interface issues,
effective control of overall ACP costs, timely
completion of the projects and carrying out
approved ACP policies and procedures. In
terms of provision of general project support
services, NAPCO was responsible for overall

Fig. 8. West Kowloon
expressway

Fig. 9. Kowloon portal
of Western Harbour
Crossing

*Fig. 10. Central reclamation*

*Fig. 11. Olympic station on airport railway*

co-ordination of project insurance programme, services and importation of labour.

- Works departments including HyD, the Territory Development Department, Civil Engineering Department, Architectural Services Department and Water Supplies Department were responsible for implementing individual ACP projects, except the airport, the airport railway and the Western Harbour Crossing.

- Non-government agencies such as AA, MTRC, and the Western Harbour Crossing franchisee were given the responsibility and authority, through enabling legislation and agreements with the government, for planning, financing, and implementing the airport, the airport railway and the Western Harbour Crossing, respectively. They also participated in the government's overall reporting system, and provided the project management information required by NAPCO for monitoring purposes, including cost controls, programme updates and progress reports.

## Project management approach and methodologies

A comprehensive integrated programme/project control approach and system was devised and implemented by NAPCO for the ACP. Details of the approach and system, including controls, administrative and reporting requirements, were contained in a set of procedures and standards as the basis for consistent project management, control and reporting across the ACP for works agents to follow.

The overall management methodology was 'top-down/bottom-up'. ACP programme-level objectives

Fig. 12. ACP manage-
ment structure

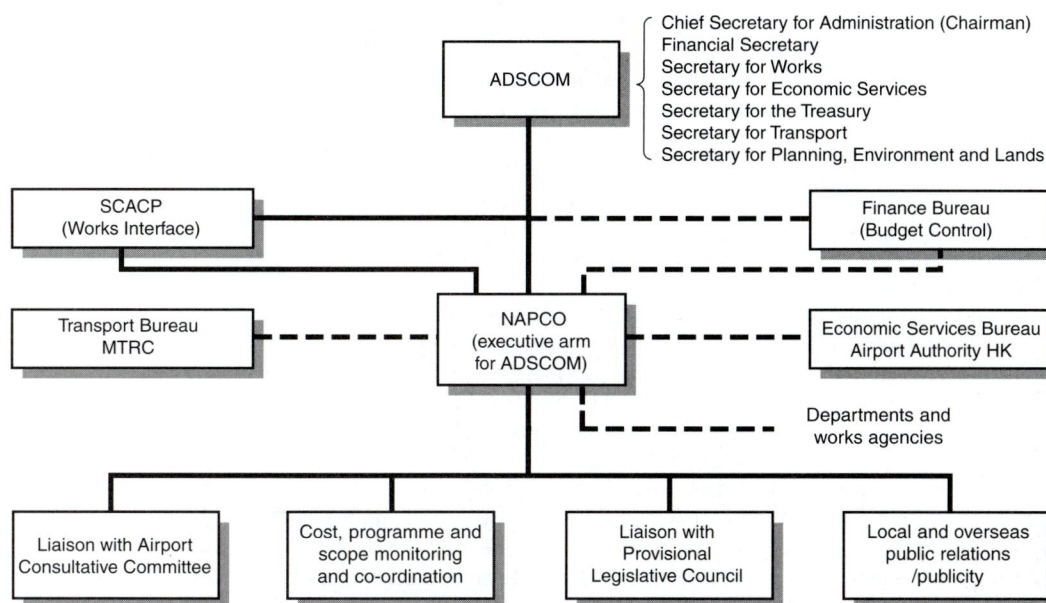

Fig. 12. ACP management structure

were established by NAPCO in conjunction with the works agents, endorsed by ADSCOM and adopted as baseline targets at the project and contract levels. Progress and status details were analysed against the targets and progressively summarized upward through the works agents to NAPCO.

### Baseline planning and implementation

The basic programme and cost control tool was the formal ACP baseline implementation plan approved by ADSCOM, which set out a clear definition of the scope and budget of each ACP project as well as master milestone and interface schedules to execute them. The ACP baseline implementation plan and its annual update provided a comprehensive road map complete with interim milestones, interface handover dates and completion targets. It was the basic frame of reference by which NAPCO monitored the projects, detected and resolved problems, and was the prerequisite for programme-level change control.

### ACP programme controls

Given the constraints, complexities and interdependencies of the ACP, as well as the fact that interim delays would often have cost implications, programme control was the centrepiece of ACP project management. An integrated hierarchy of programme was developed by NAPCO for the ACP, in line with the top-down/bottom-up management methodology. This hierarchy, which was implemented at all levels using a computerized critical-path method scheduling system, extended from detailed contract works programmes to project, interface, master schedule and executive summary programmes. The ACP programme hierarchy at the lowest level of detail encompassed some 225 construction contracts and 350 000 activities at peak. Approximately 25 000 activities and 1100 major interfaces were maintained in the master programme.

The status of each programme was reported monthly or more frequently as was required by NAPCO against both the original control and current control baselines to indicate progress to date. Potential interface conflicts and schedule problem areas were flagged and addressed in critical item action reports for resolution via NAPCO. The project programmes took account of other key activities such as land acquisition and funding approval.

### ACP cost controls

The cash limiting cost control scheme, introduced in early 1992, was one of the primary methods of cost control for the ACP. It required that funding requests for government ACP projects be submitted through NAPCO and the Finance Bureau, in order that government could ensure that such requests were within the ACP baseline budget.

Lump sum, fixed price contracts were adopted to the greatest extent practicable to enhance certainty of final project cost. The money-of-the-day concept was also introduced, requiring estimated project cash flows be adjusted with predicted price escalation factors to allow for the probable effects of inflation on construction costs, to facilitate monitoring and control of cost within approved budgetary ceilings.

As the ACP entered full-phase construction, cost estimates within the baseline budget were replaced by contract award values, and control was increasingly extended to a contract level of detail. As contract works progressed, changes, variations and claims were closely tracked.

Two levels of contingency—contract contingency and project contingency—were allowed for in the project planning stage to handle variations. Contract contingency was controlled by works agents and was used to cater for design evolution, and claims and variations within the terms of the contracts. Project contingency was used primarily for programme-related changes, to counter the effects of potential delay and to cater for programme adjustments necessary to maintain progress to meet programme objectives. Works agents had to approach NAPCO for all changes which may have required project contingency, and changes exceeding HK$ 10 million had to be approved by ADSCOM.

The government undertook an owner controlled insurance programme to cover all the government funded contracts to achieve economy of scale and to eliminate any gaps and counterclaims amongst ACP contracts.

*Tendering of ACP contracts*

In line with Hong Kong's traditional free trade and 'open door' policy, international and local construction companies, suppliers and consultants were invited to tender for ACP contracts under a fair and open system, providing a level playing field to all participants irrespective of country of origin.

Tendering normally had to go through a prequalification process. Prequalification submissions from prospective tenderers were analysed, evaluated and assessed against predetermined selection criteria on experience, technical expertise and financial capability. Prequalified tenderers were then selected and invited to submit tenders. Contract award was strictly on the basis of compliance with tender specifications, cost effectiveness and value for money.

As at the end of January 1998, 182 major ACP contracts totalling HK$ 96·4 billion were awarded. Contractors from the following 18 countries were awarded contracts : China (31%) (Hong Kong (23%) and mainland China (8%)), Japan (26%), UK (16%), Holland (6%), France (5%), Belgium (3%), New Zealand (3%), Australia (2%), US (2%), Spain (2%), Germany (2%), Italy, South Africa, Austria, Norway, Portugal, Sweden and Denmark.

*ACP conditions of contract*

A new set of ACP conditions of contract was promulgated by NAPCO to tighten up cost and programme control. Significant reallocations of risks was also introduced. The employer, acting through the engineer, had more extensive power of control than the normal standard general conditions used in non-ACP government contracts, primarily to obtain certainty of cost and to help ensure that potential changes and claims were identified and resolved in an expeditious manner with minimum schedule impact. The important aspects of the ACP conditions of contract were as follows.

- Provision for lump sum, fixed-price form of contracts and a owner-controlled insurance programme.
- Provision for the employer, through the engineer, to make variations (additions or deletions of works) to benefit timely completion of

other ACP contracts.
- Employer's ability, through the engineer, to order acceleration of work and to order contractors to recover their delays at no cost to the employer.
- Provision for employer-referable decisions relating primarily to extensions of time and additional payment to allow a greater degree of control over the actions of the engineer on these matters.
- Stringent provision for claim notification (which could result in rejection of claims if not observed by contractors) so that the employer was informed of events likely to be disruptive to the programme and/or had cost implication at an early stage.
- Introduction of a tiered disputes resolution process (mediation, adjudication and arbitration, with mediation mandatory) to help achieve significant time and cost savings when disputes arose.

*Disputes resolution and claims settlement*

To supplement the ACP conditions of contract, NAPCO promulgated procedures and guidelines to the works agents to deal with the evaluation, assessment, settlement and resolution of claims and disputes, and an integrated monitoring system was created for management of claims under the government projects.

The procedures provided for the establishment of claims advisory panels to handle difficult claims that could adversely affect project cost and time. The panels were staffed by representatives from the works agents, the engineers and NAPCO. Such provision enabled the government to focus attention on problem areas, assess the risks attached to claims and take proactive steps to deal with them.

Apart from mediation, adjudication and arbitration, negotiation was also a means of claim resolution. The objective of such extra-contractual negotiation was to achieve management benefits with identifiable cost or programme savings and to speed up the claim settlement process.

## Present status of the ACP

The government has implemented successfully the above methodologies and controls on the ACP. As at end of July 1997, progress on all fronts has been satisfactory. The seven supporting government infrastructure projects and the Western Harbour Crossing have been completed, and the road transport corridor from Central Hong Kong to the new airport has been opened for public use. The two reclamations in west Kowloon and the Central district are ready for development and population intake at Tung Chung has commenced. The new airport at Chek Lap Kok and the airport railway was opened in July 1998. The whole of the ACP can confidently be completed on time and within budget.

CIVIL ENGINEERING
DESIGN

*Proc. Instn Civ. Engrs,
Civ. Engng, Hong Kong
international Airport,
Part 1: airport,* 1998,
**126**, 15-34

*Paper 11522*

*Written discussion closes
15 November 1998*

# Hong Kong International Airport—civil engineering design

*G. W. Plant,* PhD, BEng, CEng, FICE, FHKIE, MASCE, MSAICE, Pr.Eng, *and*
*D. E. Oakervee,* CEng, FICE

Hong Kong's new airport covers an area of 1248 ha, of which three quarters has been reclaimed from the sea. This paper describes the civil engineering design for the HK$ 50 billion project including the reclamation, runways, 550 000 m$^2$ passenger terminal building, ground transportation centre and supporting infrastructure. It covers the early site selection and the factors affecting the layout of the airport. It details the leading role taken by the Authority in the prediction of reclamation settlement and how this was accounted for in the design and construction of the works. It concludes with a brief description of the wide-ranging environmental initiatives which were applied at all stages of this massive project.

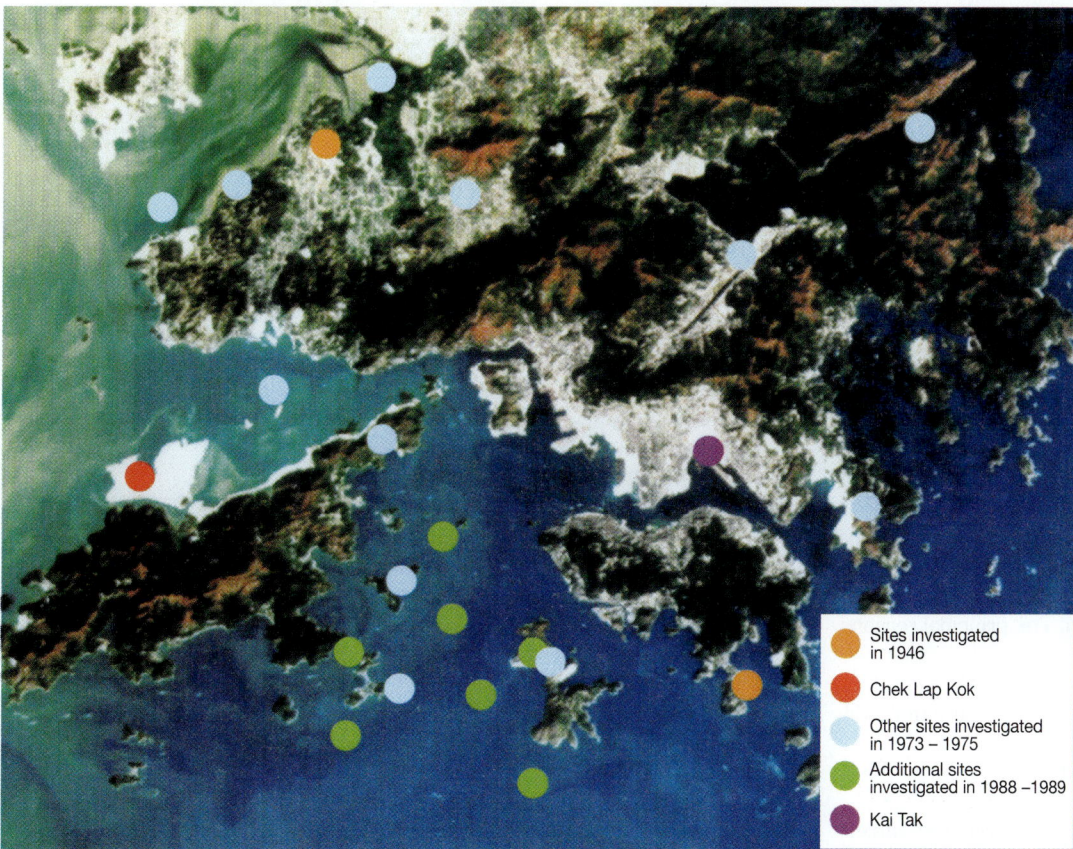

The new Hong Kong International Airport has been built on a largely man-made island at Chek Lap Kok and was opened in July 1998. The airport is a replacement for the single runway at the existing international airport at Kai Tak, situated on the Kowloon Peninsula, and which has played such a pivotal role in Hong Kong's development over the past 50 years.

*Fig. 1.Replacement airport sites investigated 1946–1989 overlain on Landsat image of November 1994 (courtesy of Civil Engineering Department, Hong Kong government)*

*Graham Plant is head of engineering at the Hong Kong Airport Authority*

*Douglas Oakervee is project director at the Hong Kong Airport Authority*

*Fig. 2. Chek Lap Kok Island*

Alternative sites for a replacement airport were considered as far back as 1946 (Fig. 1).[1] The two sites examined at that time were rejected for aeronautical reasons and high cost. To meet the steadily increasing demand at Kai Tak a new runway (2542 m long), parallel taxiway and passenger terminal were planned in the 1950s and opened in 1958. With the introduction of heavier, wide-bodied aircraft the runway was subsequently extended to 3392 m in 1975. Also in the 1970s it was realized that the single runway at Kai Tak together with constrained operating hours (due to noise over Kowloon) would have a finite capacity. A new airport was needed with dual runways, preferably aligned along an easterly axis into the prevailing winds; and good land access would be essential. Thus the first serious search for alternative sites was made between 1973 and 1975.[2] The study concluded that Chek Lap Kok (Fig. 2) would be the best choice at an estimated cost of HK$ 3·6 billion (1974 prices), including the cost of road access on north Lantau. The principal reasons for its selection were that

- all aeronautical criteria would be met for a 24-hour a day operation
- the engineering feasibility for site formation was established
- the programme for the works was less uncertain than other sites
- the environmental impact during construction and for operations was acceptable
- an airport at Chek Lap Kok was compatible with other proposed developments on Lantau Island and also with port activities in the western harbour.

In 1978 studies undertaken by the government looked at the development of an airport at Chek Lap Kok together with land development potential on north Lantau with a coastal highway and a fixed link crossing to Tsing Yi.

Master planning and financing studies completed in 1982[3] established in some detail the feasibility of building the two-runway airport at Chek Lap Kok within a seven-year period at a total cost of HK$ 15·3 billion (mid-1982 prices). However, due to a downturn in the global economy and uncertainties arising from concurrent Sino-British negotiations on Hong Kong's future, the project was deferred.

However, while planning came to a standstill the growth of air traffic through Kai Tak continued unabated. Between 1980 and 1990 alone, annual passenger numbers rose from 8 million to 20 million and the volume of air cargo passing through the airport trebled.

By 1989, the need for a new airport could no longer be ignored and once again planners focused on Chek Lap Kok, but in a new context. This time the airport was to be studied as part of a combined port and airport development strategy and included a review of alternative sites to Chek Lap Kok.

In October 1989, the Hong Kong government announced the adoption of a strategy based on a new airport at Chek Lap Kok, and a network of supporting road and rail infrastructure to be known as the Airport Core Programme.

Chek Lap Kok was chosen as the airport location ahead of other site options because it met a number of basic criteria. It offered greater economic benefits and could be built and made operational earlier than other considered sites. It also provided greater scope for an integrated and viable transport infrastructure network, the potential for urban development in the form of Hong Kong's ninth new town and the ability to integrate future port and airport development. The government emphasized that the airport would develop incrementally with expansion capacity for phased development well into the next century.

The Provisional Airport Authority (PAA) was established in April 1990 as a statutory organization to plan, design and construct the new airport. The new airport master plan[4] was also initiated in July 1990. In September 1991 a memorandum of understanding was signed by the governments of the UK and the People's Republic of China, confirming joint support for the new airport and related core projects. The cost of the PAA's facilities at the airport in money-of-the-day or 'out-turn prices' is HK$ 49.8 billion. There followed almost four years of discussion between the two governments on the detailed financing arrangements for the projects. During this time, the Hong Kong government sought to maintain progress on the airport in accordance with the requirements of the memorandum by, at various stages, providing interest-free cash advances to the PAA to enable critical works on the project to proceed on a 'step-by-step' basis. In July 1995 Hong Kong's legislative council passed the Airport Authority Bill which enabled the PAA to be reconstituted as the Airport Authority (AA), empowering it to provide, develop, operate and maintain the new airport. On 29 November 1995 the airport committee of the Sino-British joint liaison group signed agreed minutes on the membership of the AA, and this paved the way for establishment of the AA on 1 December 1995.

The above were key events for the AA and had a fundamental influence in the contracting of the design and construction of the airport.

## Master plan

The overall objective of the new airport master plan (NAMP) study was defined as 'the preparation of a comprehensive and environmentally acceptable scheme for the planning and implementation of an operationally safe and efficient new airport at Chek Lap Kok, covering progressive development into a two-runway airport operating 24 hours a day'.[4]

The key assumptions made for the NAMP, completed in December 1991, which influenced the planning of the airport facilities included the following.

- Kai Tak Airport would close when the new airport opened.
- All passengers using the new airport would be regarded as international, requiring them to be processed through immigration and customs facilities.
- The master plan would include a multi-modal surface transport system with a high priority given to rail.

*Table 1. Air passenger and cargo demand forecasts*

| Year | Passengers | | Cargo: tonnes | |
|------|------------|------------|----------------|----------------|
| | 1991 forecast | 1994 forecast | 1991 forecast | 1994 forecast |
| 1998 | 29.5 million | 33.9 million | 1.2 million | 1.7 million |
| 2001 | 33.2 million | 40.7 million | 1.4 million | 2.1 million |
| 2010 | 44.7 million | 65.9 million | 2.3 million | 3.4 million |

- Opportunities for privatization would be maximized.

There were three main workstreams undertaken for the NAMP, namely planning, civil engineering and environmental impact assessment, which were carried out in parallel.

## Air traffic demand

Air traffic studies carried out as part of the NAMP determined the overall physical size of the airport and the size of the facilities. The air traffic forecasting process employed has been described by Oakervee.[5] In essence, the predicted per capita gross domestic product was used as the primary variable to determine growth.

Such has been the economic development in the Asia–Pacific region that in 1994 a new air traffic demand forecast covering passengers and cargo was commissioned by the AA, as air traffic growth in the three years since completing the 1991 master plan had been significantly greater than forecast and reported previously.[5] The 1991 and 1994 forecast figures are given in Table 1.

Since 1987 annual passenger traffic has grown by an average of 9.9%[6] and in 1996, despite its limitations, Kai Tak had an annual throughput of 30.2 million passengers, making it the third busiest airport in the world in terms of international passengers.

Air cargo tonnage has grown by an average of 11.0% since 1987. Based on preliminary figures from the Airports Council International, Kai Tak was the world's busiest in terms of international cargo in 1996 with a throughput of 1.56 Mt.

The forecast of passenger and cargo aircraft movements was strongly influenced by the relatively high predominance of Boeing 747 and other wide bodied aircraft. Both the 1991 and 1994 forecasts for all aircraft movements are given in Table 2.

## The airport plan

Air traffic demand and airspace constraints dictated the layout of the airfield at Chek Lap Kok which covers a total area of 1248 ha. The major planning recommendations in the NAMP were that runway centre-lines should be separated by 1525 m which would allow independent runway operations in accordance with the recommendations of the International Civil Aviation Organization (ICAO); the airport would be

*Table 2. Forecasts of all aircraft movements*

| Year | Movements | |
|------|-----------|-----------|
| | 1991 forecast | 1994 forecast |
| 1998 | 160 000 | 201 000 |
| 2001 | 177 000 | 241 100 |
| 2010 | 229 300 | 369 900 |

designed for a configuration of two parallel runways bearing 070°/250° N each 3800 m long and 60 m wide; the passenger terminal complex would be located between the runways; and the rail and primary road access would be along the eastern edge of the airport island.

The airport is a phased development and as such has been planned to open with a design capacity of 35 million passengers and 3 Mt of cargo annually. Initially, has a single runway, one passenger processing building, the major part of the Y-shaped concourse with 38 frontal gates (connected to the terminal by airbridges) and 27 remote aircraft stands. Within months of the airport opening, the second runway will become operational, the remainder of the Y-concourse will be constructed and a further 10 frontal gates will be added. Appendix 1 gives a comparison between facilities at the existing airport at Kai Tak and those at the new airport at Chek Lap Kok. Ultimately the airport will have a designed capacity of about 87 million passengers and 9 Mt of cargo annually.

Airport planning is a continuous process and following completion of the NAMP the AA made further design refinements and revisions in the light of changing data and circumstances. The significant changes (Fig. 3) were that: the southern runway was moved 360 m to the west to reduce reclamation costs and to preserve the eastern coast of Chek Lap Kok Island; the passenger terminal building was moved 190 m westward so that more of the building's foundation work was located on Chek Lap Kok Island to reduce construction costs; and the vehicular tunnels were re-aligned.

The key features of the airport site plan (Fig. 4) are two parallel runways, each 3800 m long, and 35 km of associated taxiways; a passenger terminal building comprising a Y-shaped concourse and processing terminal connected by way of an underground automated people mover and linked to the ground transportation centre (GTC); a high-speed rail and road network serving the GTC; aircraft base maintenance facilities at the western end of the airfield; airport support facilities including air cargo, catering and an aviation fuel tank farm located south of the runways; various government facilities including the air traffic control tower and centre, fire stations, sea rescue stations, airmail centre, a police station and a facility for the government flying service. Commercial developments are also under construction, including a 1100-room hotel, a 1750-space multi-storey car park, a freight forwarding centre, headquarters and crew accommodation for Cathay Pacific and headquarters for Dragonair/CNAC.

## Design contracts

The NAMP identified general requirements for all the airport facilities and included outline scheme designs for the passenger terminal building, the airfield, the infrastructure and utilities. The first major detailed design contract for the

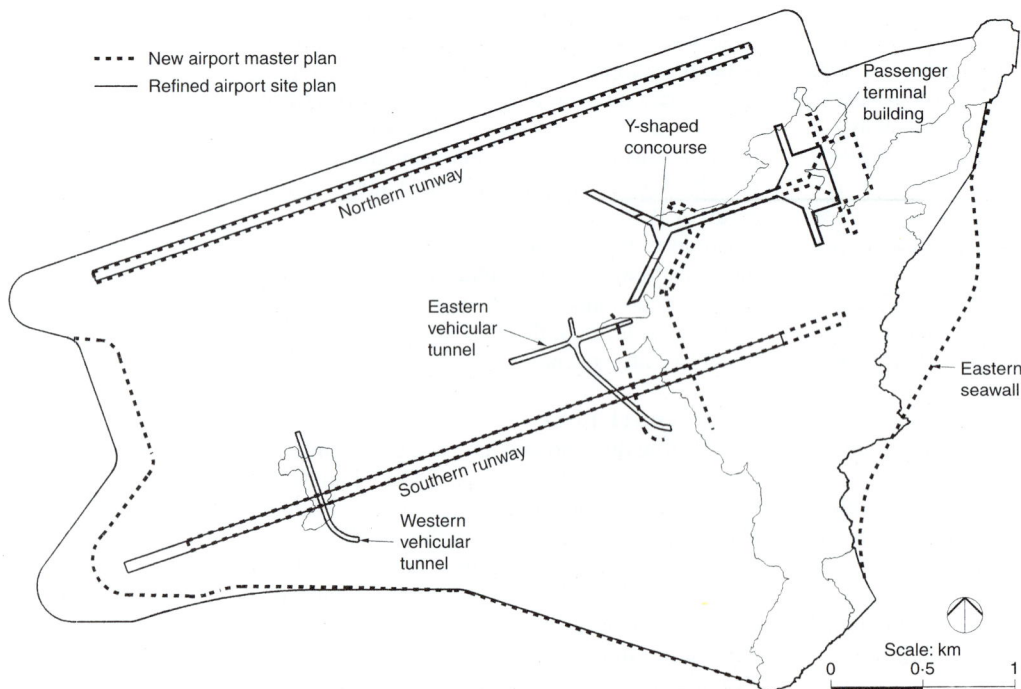

*Fig. 3. Key changes to the airport site plan*

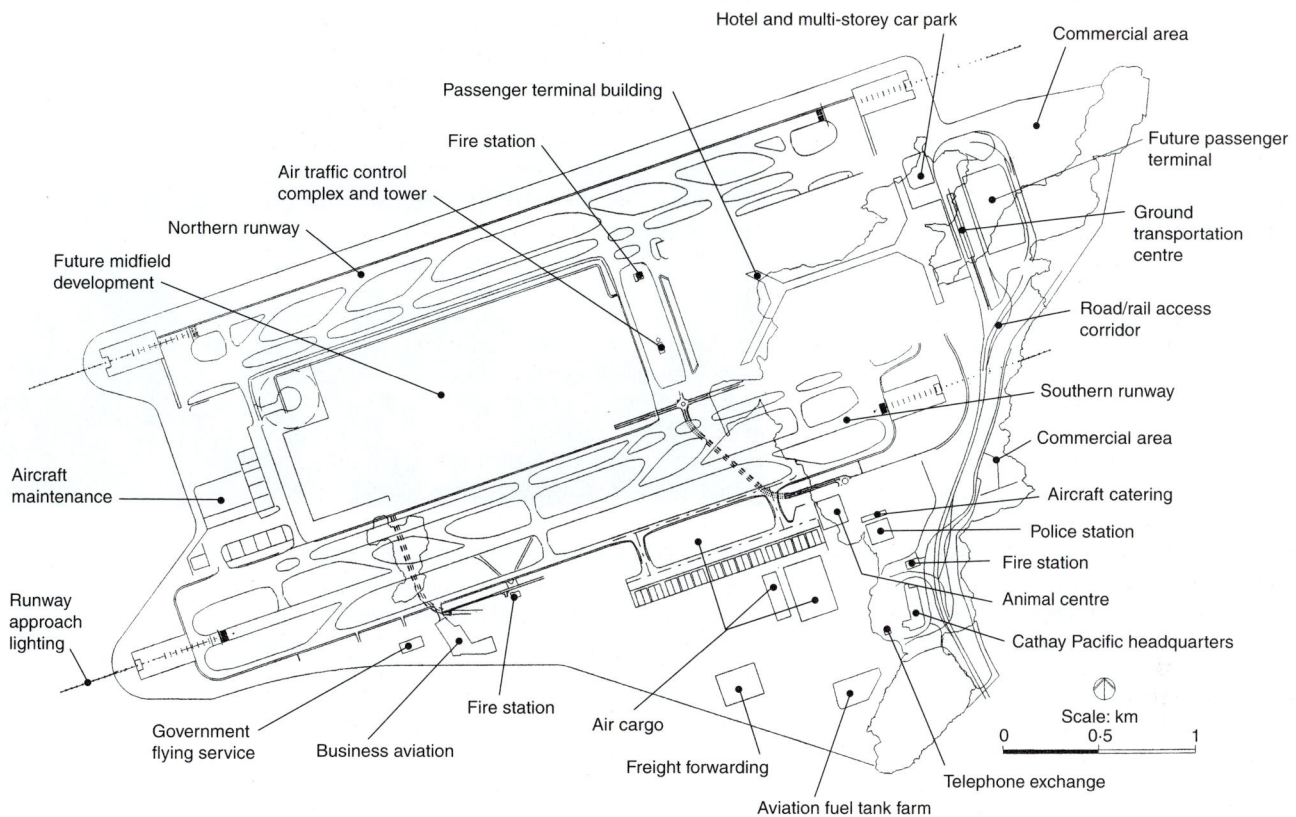

*Fig. 4. Airport site plan*

airport itself was awarded in March 1992 for the passenger terminal building. For the remainder of the works a large number (30) of contracts were let between 1993 and 1996 to design consultants. A full list of the design contracts is given at Appendix 2. The reasons for this strategy arose from the limited funding available at the time which required all design work to be initiated on a 'step-by-step' basis. Also, the AA's board wished to have participation by a number of designers, each expert in specific areas—airfield pavements, ground lighting, stormwater drainage, utilities, electrical and mechanical works, roads, bridges, ancillary buildings and landscaping.

The allocation of funds on a step-by-step basis led to interface problems such as the stormwater design being awarded before the airfield pavements package and the utilities design before the roads package. This inevitably created complications and led to an iterative design process. The consequences of a large number of design interfaces and design work being undertaken out of sequence were felt throughout the design and construction phases. In addition, the fluid nature of the contracting strategy for the construction works which arose from the uncertainty over future funding meant that the completed designs from several designers had to be split and reconsolidated to suit each works contract.

Design co-ordination meetings, chaired by the AA and attended by all design managers and representatives from all designers, were held periodi-

cally for co-ordination of all designs. The authors would like to acknowledge the cooperation of designers in dealing with the interfaces and assisting in the co-ordination process. The AA also held quarterly review meetings with all design firms to review progress at a strategic level, which proved beneficial to all concerned.

An example of the AA's initiatives with regard to co-ordination was with respect to the provisions necessary to account for settlement of the airport platform and this is described later in this paper under the heading *Design and construction levels*.

## The airport platform

To accommodate all the facilities and allow for phased development of the airport required the creation of 1248 ha of land (Fig. 5). Twenty-five per cent of the airport platform is made up of the former islands of Chek Lap Kok and Lam Chau, which have been excavated to a level of about +6 m PD (principal datum). The remainder of the platform is land which has been reclaimed from the sea as part of a major site preparation contract which began in December 1992, the construction details of which are briefly described in a related paper.[7] A full description of the design, construction and performance of the site preparation works is given in reference 1.

## Design and performance

Chek Lap Kok and Lam Chau Islands are formed predominantly of granite. Completely decomposed granite (soil) covered the surface of both islands, generally 0·5 to 3 m thick but up to 20 m thick in the vicinity of faults. The offshore geology comprises a complex sequence of interbedded clays and sands. This is overlain by recent (Holocene) marine clays and is underlain by granite bedrock in various states of weathering.[8]

The mean sea level at the site is +1·3 m PD and the average water depth prior to reclamation was typically 5 m, increasing to 15 m in places. The average thickness of the recent soft marine mud was 8 m. The total thickness of the underlying alluvial clays and sands varies from about 10 m to 30 m.

The decision to adopt a dredged reclamation was made at an early stage in the design of the platform because the results from a test embankment constructed in 1982[9] indicated that if the soft marine mud were to be left in place then an additional 2 m to 3 m of settlement would occur. Also, that wickdrains would be required at very close centres to be effective within the airport's fast-track programme. The selected dredging depth was based on the results of 3200 cone penetration tests carried out on a 50–100 m grid across the site combined with seismic traverses. The dredge levels within the reclamation (i.e. excluding seawall trenches) were defined by removing mud with a net cone penetration test (CPT) tip resistance of less than 500 kPa. The dredge level was specified with a tolerance of –0 m, +1 m.

The seawalls relied upon the strength of the founding materials for stability. For the purposes of assessing short-term stability a minimum undrained shear strength of 35 kPa was adopted. It was found that, in general, this could be met with a net CPT tip resistance of 700 kPa. The dredge level tolerance was specified as +0 m, –1 m to minimize the possibility of mud remaining beneath the seawalls. In all, dredgers removed 68·8 Mm³ of soft marine mud overlying the seabed within the airport footprint to depths up to –29 m PD and transported it to designated dumping grounds within Hong Kong waters.

The total fill requirement was 197 Mm³, of which 108 Mm³ was obtained from the islands of Chek Lap Kok and Lam Chau, 6·6 Mm³ from the levelling of the two Brothers Islands to the east of the airport and 7·3 Mm³ from other ACP contracts. The balance was sand imported from marine borrow areas within Hong Kong waters. The reclamation is surrounded by 13 km of seawall, the majority of which is armoured sloping wall.

Three main material types, A, B, and C, were specified for use as general reclamation fill. A simplified version of the fill allocation plan is shown in Fig. 6. A number of other fill types were adopted for specific uses such as the seawalls. Type A is a well-graded hard durable rock fill with a maximum size of 2 m (Fig. 7) and was reserved for use below the southern runway for its resistance to liquefaction potential. Type B is a well-graded fill from excavation; in practice,

*Fig. 5. The airport platform*

essentially completely decomposed granite (CDG) with a maximum size of 300 mm and was used in areas where future excavation of the cut-and-cover airfield tunnels were envisaged or in areas where structures were to be founded on piles. Type C is sand from marine sources. A fourth material type, designated type A/B, was agreed during the course of the site preparation contract for use in general fill areas. In practice the material is a rockfill with a small but variable content of CDG. The upper 2 to 3 m of the reclamation comprises a capping layer of either type B or C fill. This permitted subgrade preparation for the airfield pavements and eased excavation for drainage and services trenches as well as final shaping of landscaped areas.

The question of settlement is a key issue for reclamations and was considered from an early stage. To estimate the amount and rate of settlement required, a thorough understanding of the geology and the engineering behaviour of the soils was needed. Site investigations included the reclamation site and the potential sources of fill for reclamation. About 650 marine borings, 170 land borings and 1100 km of seismic traverses were carried out on the airport footprint during the detailed design stage. These investigations were carried out by different organizations over many years and a comprehensive geotechnical digital database was established by the AA. This database links all ground investigation data, stratigraphic layers, laboratory tests and instrumentation results. The database is also linked to engineering drawing applications and has proved a valuable tool in planning, design and analysis.

The majority of the reclamation was constructed by end-tipping of rockfill and bottom dumping or hydraulic placement of marine sand with no subsequent ground treatment. Therefore, where necessary, limited areas of the reclamation have been treated generally by either surcharging or vibrocompaction (Fig. 8). Surcharging, principally using 8 Mt of granite rockfill retained from quar-

*Fig. 6. Simplified fill
allocation plan*

Type A

Type B

Type A/B

Type C to −9 mPD
Type A/B above −9 mPD

Type C

Type B to −3 mPD
Type C above −3 mPD

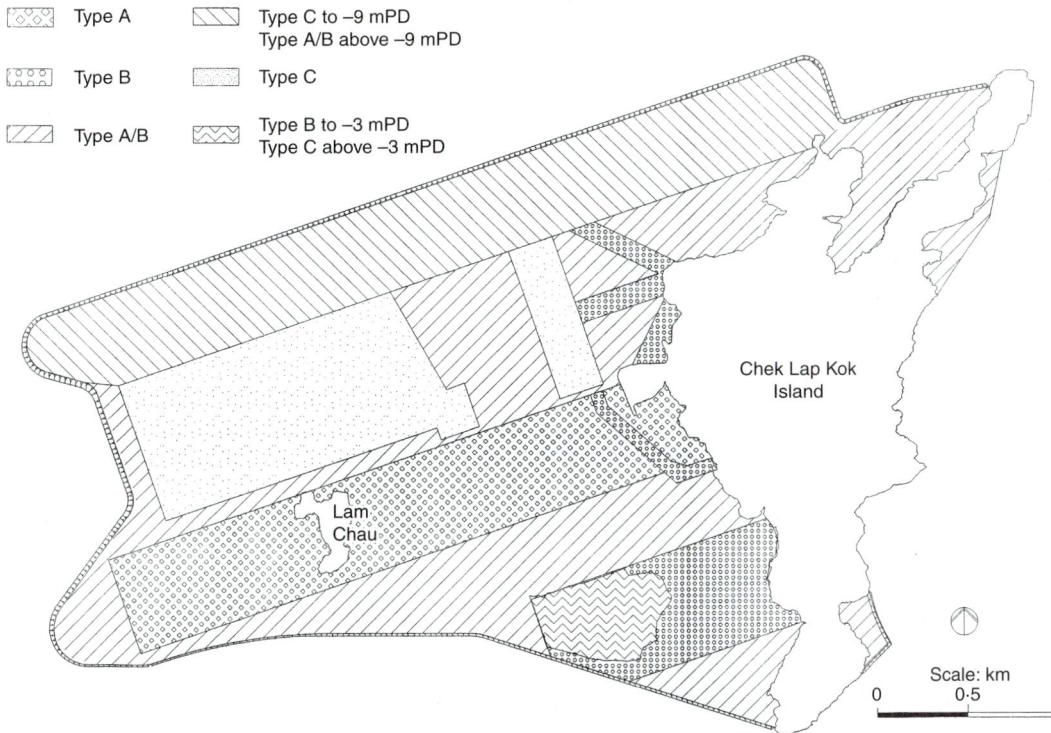

Chek Lap Kok
Island

Lam
Chau

Scale: km

0      0·5      1

*Fig. 7. Grading
envelopes for fill types
A, B, A/B and C*

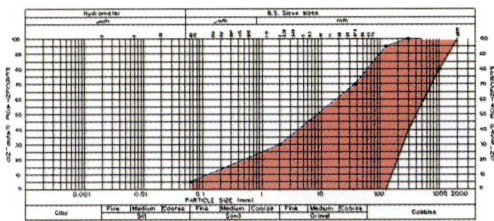

Fill type A

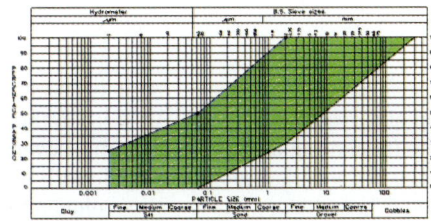

Fill type B

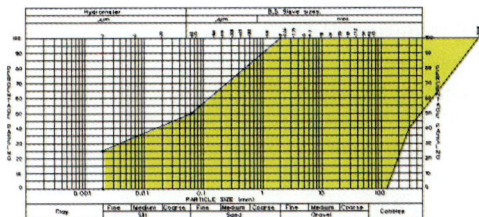

Fill type A/B

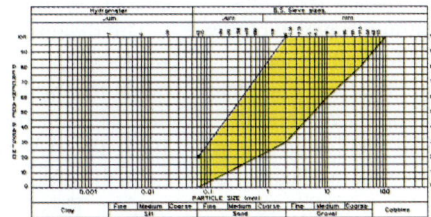

Fill type C

rying Chek Lap Kok Island and subsequently crushed for aggregates, was designed to accelerate primary consolidation of the alluvial clays and also to improve the creep characteristics of the rockfills. Vibrocompaction of 11 Mm³ of the sandfill was carried out over some 60 ha of land being developed under the current construction phase. The treatment was required to reduce creep settlement of the sandfill and to reduce the potential for vibration induced settlement during follow-on piling and other construction activities. The acceptance criterion for the vibrocompaction works was a minimum CPT value of either 8 MPa or 15 MPa, depending on the intended land use.

An integral part of the reclamation design was the installation and monitoring of surface and sub-surface geotechnical instruments to assess the performance of the reclamation and to allow predictions of future settlement to be refined. The sub-surface instruments comprised clusters of piezometers installed around a central extensometer at approximately 500 m centres. A grid of surface survey markers was constructed across the reclamation at approximately 200 m centres to provide a fuller picture of the surface settlement of the reclamation.

Early predictions of settlement were made using the results of *in situ* and laboratory tests. As more

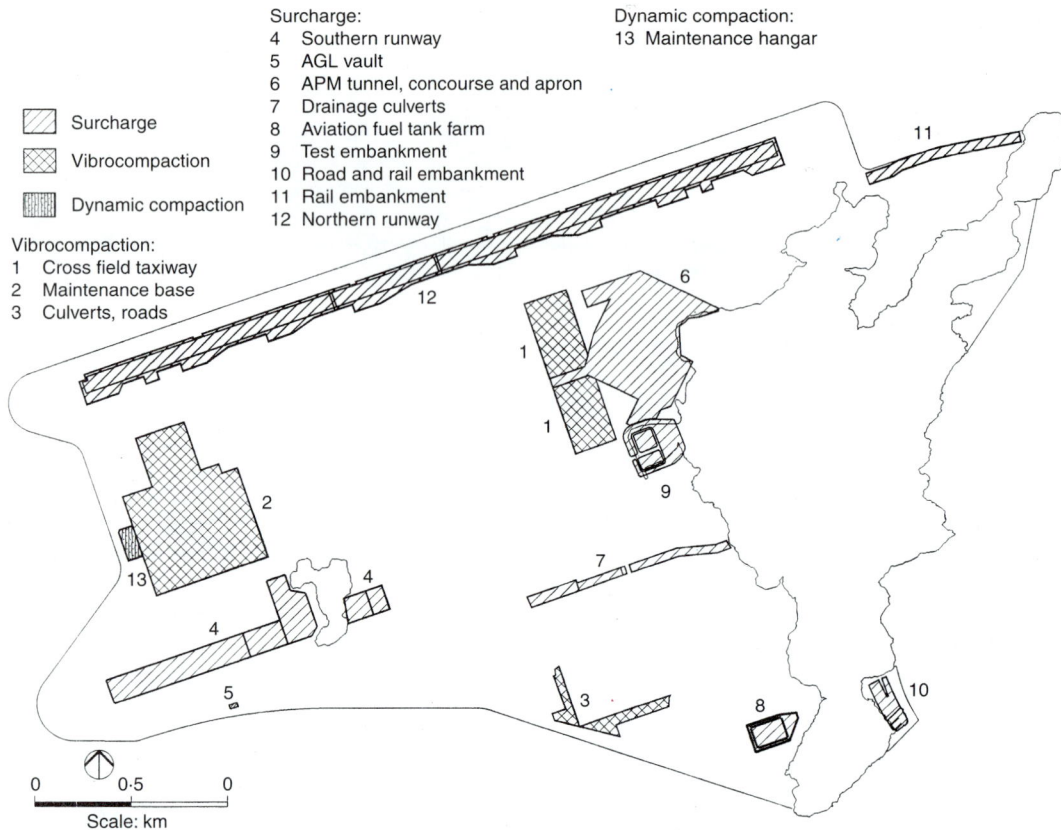

Surcharge:
4 Southern runway
5 AGL vault
6 APM tunnel, concourse and apron
7 Drainage culverts
8 Aviation fuel tank farm
9 Test embankment
10 Road and rail embankment
11 Rail embankment
12 Northern runway

Dynamic compaction:
13 Maintenance hangar

Surcharge
Vibrocompaction
Dynamic compaction

Vibrocompaction:
1 Cross field taxiway
2 Maintenance base
3 Culverts, roads

Scale: km
0    0·5    0

*Fig. 8. Ground treatment areas*

monitoring data have become available, the prediction method has changed to an observational approach making full use of the monitoring data. Fuller details of the performance of the reclamation are given in a companion paper. [10] In summary, the remaining or residual settlement from January 1997 to 2040 is typically in the range 200 mm to 500 mm. At the present time (August 1997) the primary consolidation of the alluvial clay is typically 90% complete and creep of the reclamation fills now typically comprises about 50% of the residual settlement.

The airport platform was substantially completed in 37 months and involved the movement of about 360 Mm$^3$ of material. The rapid pace of construction was assisted by a practical application of engineering requirements and a carefully programmed fill allocation plan taking best advantage of the various fill types available. This minimized the requirements for fill processing and ensured that the fill placed in a particular area met the specific engineering needs of the future planned construction. This approach also minimized the requirement for ground improvement thereby reducing both cost and time.

## Design and construction levels

The effect of the platform settlement coupled with the different times at which works were due to be constructed influenced the amount of overbuild necessary to achieve design levels in the long term.

Although each designer made his own settle-

ment assessment it soon became clear that the AA were in the best position to address this issue from their overall knowledge of the platform behaviour. It was also known that contractors were reluctant to take on risk associated with the settlement of the platform. The AA therefore took on responsibility for settlement during construction of the follow-on works and established the concept of installation levels.

The designers were required to design their works such that the facilities would remain operational throughout the airport's life and to standardize their designs for a fixed date of January 1997. The AA then provided contractors with rates of settlement for various zones of the reclamation. This information was derived from time–settlement curves prepared for many points on the platform. The construction period for the works represented a small window within the overall time–settlement curve (Fig. 9) and was therefore approximated to a straight line. The AA issued settlement rates expressed as mm per month and the contractor then determined his installation levels from this information and from his own programme for the works by taking the linear rate of settlement multiplied by the number of months prior to January 1997 and adding this amount to the design levels.

## Airfield pavements

Air traffic forecasts and airspace considerations had determined that the Civil Aviation

Table 3. Principal airfield planning dimensions (in m)

| Criterion | ICAO code E annex 14 | ICAO code F guidelines | FAA group V | FAA group VI | Design dimensions adopted |
|---|---|---|---|---|---|
| Runway centreline (CL) to taxiway CL | 182·5 | 192 | 120 | 180 | 192 |
| Runway width | 45 | 60 | 45 | 60 | 60 |
| Runway shoulder width | 7·5 | 7·5 | 10·5 | 12 | 7·5 |
| Taxiway CL to taxiway CL | 80 | 99 | 81 | 99 | 99 |
| Taxiway CL to obstacle | 47·5 | 57 | 48·5 | 59 | 57 |
| Taxilane CL to obstacle | 42·5 | 54 | 42 | 51 | 54 |
| Taxiway/taxilane width | 23 | 29 | 23 | 30 | 29 |
| Taxiway shoulder width | 10·5 | 7·5 | 10·5 | 12 | 15·5 |

Department (CAD, which is responsible for aircraft movements in the air and on the ground until the aircraft comes to rest at its assigned stand at which point the AA takes over responsibility) and the AA would need the capability to operate two runways simultaneously, 24 hours a day. The system of runways and taxiways occupy about 30% of the total airport site (Fig. 10). The 'rapid exit' taxiways shown permit aircraft to clear the runway more quickly than a 90° turn thus improving runway capacity.

At airport opening it is estimated that over 75% of departures[11] will be wide bodied aircraft, and thus all frontal stands and the majority of remote stands have been designed to accommodate ICAO designated code E aircraft (e.g. Boeing 747-400). The length of selected stands have also been safeguarded to ensure that the new large aircraft (NLA) can be accommodated.

Throughout the design phase, and even now, there has been much uncertainty about the size and introduction of NLAs and ICAO have only issued guidelines rather than a code F standard. These guidelines, together with the Federal Aviation Administration (FAA) group VI design criteria, were used as a basis for designing the geometry of the airfield. The principal planning criteria are shown in Table 3.

This paper briefly summarizes the design of the pavements only, as available space precludes discussion of the associated aspects such as aircraft approach and ground lighting, high mast lighting, preconditioned air, fixed ground power, HV ducting, communications, potable and fire water mains, aviation fuel distribution, irrigation mains, stormwater drainage, oil-interception, aircraft wash, parking aids and miscellaneous buildings.

The design aircraft for the structural design of the pavements was a Boeing 747-400 with an all-up weight of 400 t. The Boeing 777 and MD11 aircraft give similar pavement loadings. Growth versions of these aircraft such as the Boeing 747-600, MD12 and Airbus A3XX are expected to increase the all-up weight to between 550 and 770 t, but revised landing gear configuration and increased number of wheels

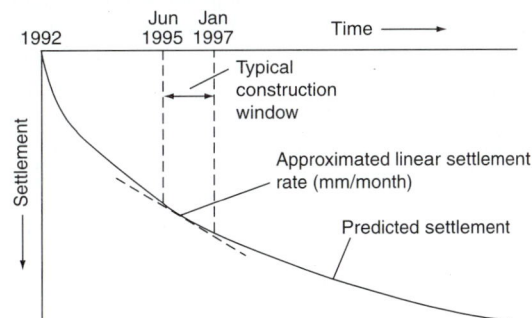

Fig. 9. Interpretation of predicted settlement to derive settlement rates

will result in similar pavement stresses. Underground structures such as culverts and tunnels have been designed to carry aircraft up to 770 t.

Surface winds at the new airport are predominantly from the east and favour a flow of operations in this direction (runways 07L and 07R) approximately 70% of the time, with the remaining 30% operating from runways 25R and 25L. However, with a 5 knot tailwind aircraft could continue to operate in the 250° direction for up to about 45% of the time. Thus, for the purposes of runway pave-

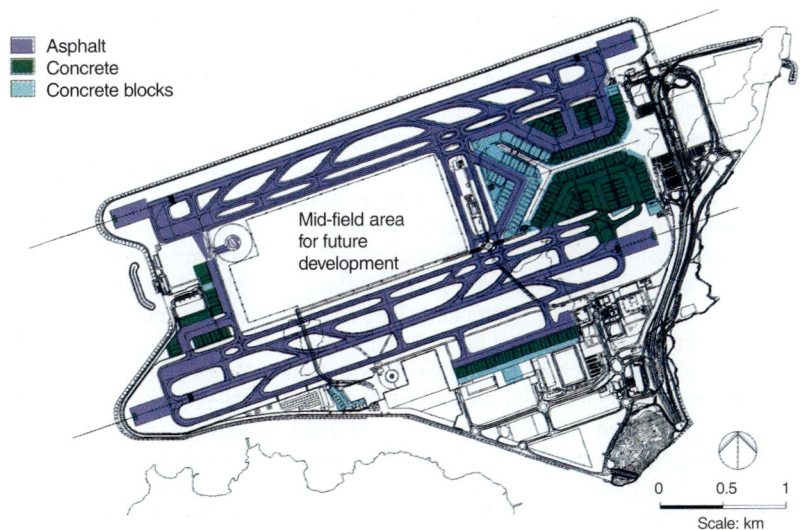

Fig. 10. Type and layout of airfield pavements

23

ment design it was assumed that the runway utilization could be 70% departures from 07L/07R and 45% from 25R/25L. A detailed traffic analysis was undertaken based on information from CAD regarding aircraft fleet mix and planned taxiing operations on the ground to establish coverages by the design aircraft over the pavements.

The principal factors which affected the choice of pavement were settlement, ride quality, fuel resistance, duration of loading and cost. Based on these design considerations, asphalt surfaced flexible pavements were selected for the runway/taxiway system on the reclamation, rigid pavement quality concrete for aprons and taxilanes on the original island and interlocking concrete blocks for apron areas subject to significant differential settlements.

The pavements have been designed using both the 1989 UK PSA *A Guide To Airfield Pavement Design And Evaluation*[10] and the US LEDFAA method following its release in 1995 by the US FAA.[12] The first is a semi-empirical method in line with those recognized by ICAO and is based on many decades of experience gained from evaluating the performance of existing pavements. The second is based on a linear elastic design approach which calculates stresses and strains in each pavement layer. It uses these stresses and strains to determine pavement thickness by an iterative process using established fatigue equations. Both of these design approaches yielded similar pavement structures.

After applying the various traffic scenarios, expected loading and settlement criteria many different pavement sections were developed across the airport. These were then rationalized into just 16 types to ease the construction process as many of the scenarios gave rise to similar thickness requirements. In all, the airfield pavement works cover 3 700 000 m$^2$.

## Asphalt pavements

The initial structural design of the asphalt surfaced pavements considered the use of a bound (cement or bitumen) base in accordance with the PSA design guide. However, unbound bases with relatively thin asphalt surfaces have provided the desired performance at many airports. These pavements require very high quality aggregates with strict grading requirements with a consistent level of compaction to prevent premature rutting.

Following the identification of suitable high quality aggregates, a pavement using crushed rock base and sub-base courses was selected on economic grounds, being significantly less costly than a pavement with a bound base. Crushed rock aggregates of the specified quality for the sub-base course were available on site, retained from quarrying Chek Lap Kok. For the base course and the Marshall asphalt layers, crushed rock with a minimum specified 10% fines value of 180 kN was imported from nearby provinces in China.

Transverse sawn grooves (6 mm x 6 mm at 32 mm centres) were specified in the runway wearing course to provide the required frictional characteristics.

The total area of both runway/taxiway systems is 2 600 000 m$^2$ and they have a design life of 20 years. Within this period overlays will be placed because of the oxidation and weathering of the asphalt which occur in tropical climates and also, in limited areas, for shape correction. Fig. 11(a) shows a cross-section of a typical asphalt pavement.

## Concrete pavements

For the aprons, concrete paving is preferred because of its ability to resist high loads from stationary aircraft and its resistance to fuel and oil spillage from aircraft or ground support vehicles. Thus conventional rigid concrete pavements have generally been adopted for the aprons. During the early phase of design both continuously reinforced and unreinforced jointed concrete pavements were considered, and the latter were selected on the basis of lower capital costs.

The rigid concrete pavements which cover an area of 700 000 m$^2$ have a design life of 45 years. [10] The majority of these pavements are on the original island; elsewhere *in situ* plate tests using a 762 mm dia. plate gave values of sub-grade reaction for the compacted fill of between 100 and 150 kPa/mm. A conservative value of 40 kPa/mm was selected for design purposes. The unreinforced pavement is typically 450 mm thick on a 175 mm dry lean concrete base (Fig. 11(b)). The specified 28-day characteristic flexural strength was 4·5 MPa. Typical bay sizes are 6 m x 6 m. Mesh reinforcement and trimmer bars were specified for irregular bays and around penetrations. Transverse joints were saw cut at approximately 10 to 15 h after concrete placement and steel dowels were provided at longitudinal joints. Load transfer across the transverse joints is by aggregate interlock at the sawn joint. All joints were sealed with a fuel-resistant sealant. Surface texture of the finished concrete was specified as a minimum of 0·7 mm using the sand patch test method.

## Concrete blocks

In apron areas where a rigid concrete pavement could not tolerate the estimated differential movements, interlocking shaped concrete blocks 80 mm thick have been adopted (Fig. 11(c)). The blocks are nominally 225 mm by 112·5 mm and have a characteristic compressive cube strength of 60 MPa. The joint spacing was specified as 2–4 mm to ensure interlock via the jointing sand. To achieve this the contractor monitored mould wear and replaced the mould when the block plan dimensions increased by 1·25 mm which occurred at about 80 000 cycles.[13] Some 16 million blocks were required to cover 400 000 m$^2$ of apron.

Beneath the southern runway the eastern tunnel has been designed to carry vehicles from the

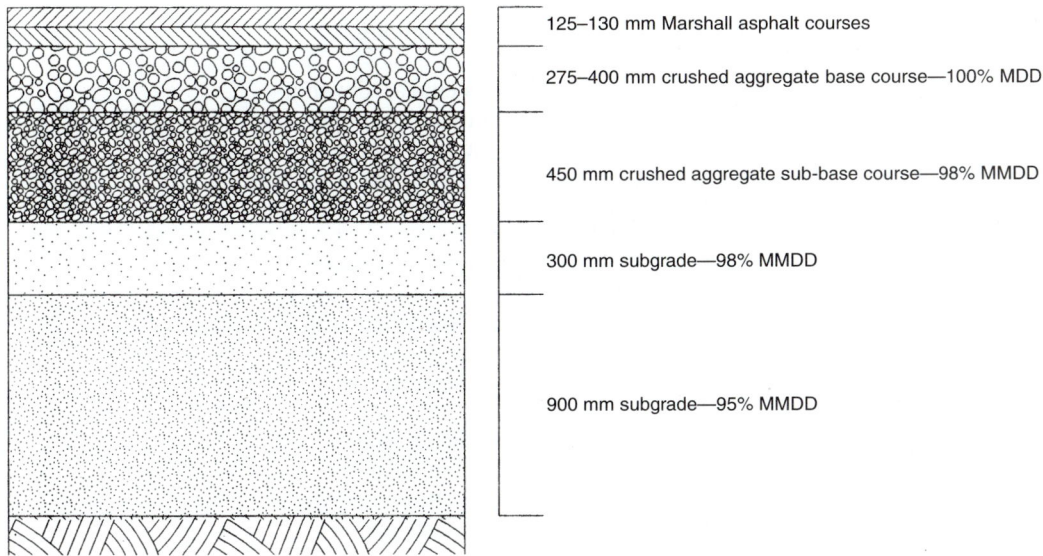

125–130 mm Marshall asphalt courses

275–400 mm crushed aggregate base course—100% MDD

450 mm crushed aggregate sub-base course—98% MMDD

300 mm subgrade—98% MMDD

900 mm subgrade—95% MMDD

(a)

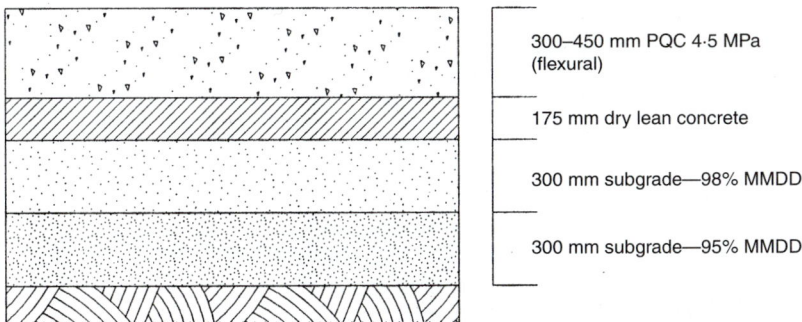

300–450 mm PQC 4·5 MPa
(flexural)

175 mm dry lean concrete

300 mm subgrade—98% MMDD

300 mm subgrade—95% MMDD

(b)

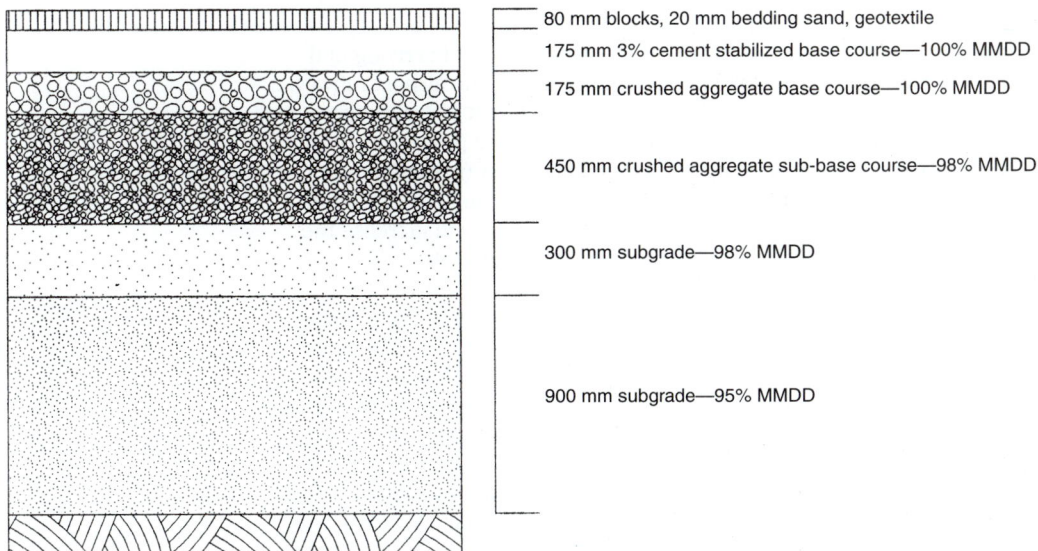

80 mm blocks, 20 mm bedding sand, geotextile
175 mm 3% cement stabilized base course—100% MMDD
175 mm crushed aggregate base course—100% MMDD

450 mm crushed aggregate sub-base course—98% MMDD

300 mm subgrade—98% MMDD

900 mm subgrade—95% MMDD

(c)

*Fig. 11. Typical pavement types (a) asphalt, (b) rigid concrete, (c) concrete blocks*

airport support facilities in the south to the passenger terminal aprons, and the western tunnel is for future access to the midfield. Both tunnels are each about 700 m long and have a utilities cell and dual carriageways to allow traffic from landside to airside without affecting operations of the southern runway. The eastern tunnel has been commissioned but the western is for utilities only at present and will be fitted out when the mid-field area is developed (Fig. 10).

## Infrastructure

In addition to the terminal building for passenger handling and the aircraft operations, the airport demands significant supporting infrastructure. This includes

(a) air cargo handling and freight forwarding facilities
(b) in-flight catering services
(c) aviation fuel storage and distribution
(d) aircraft maintenance, line maintenance and ramp handling
(e) government facilities
(f) building and E&M airport maintenance depots
(g) utilities and drainage
(h) roads, tunnels and bridges.

Items (a) to (d) are being developed by commercial franchisees, item (e) by government and items (f) to (h) by the AA.

*Franchises*

At airport opening two air cargo franchisees will have a capacity to handle at least 1·6 Mt of cargo which will quickly rise to 3 Mt. A freight forwarding facility where cargo is consolidated for direct dispatch is also being built. Three aircraft catering facilities are being built and between them will produce about 95 000 meals per day during the first year of operation.

The design, construction and operation of the aviation fuelling system has been licensed to industry specialists. Two 450 mm diameter undersea pipelines will bring jet A1 fuel on to the airport, where 15 days of storage is provided. The throughput is about 10 000 000 litres per day initially. A fuel hydrant distribution system, which transfers fuel from the tank farm to the aircraft stands for aircraft refuelling, was designed and constructed by the AA on behalf of the licensee. It is sized for busy-hour demands for the year 2010 and is expandable for further growth. An aircraft maintenance facility is under construction at the west end of the airport. The main component is a 220 m x 66 m structural steel hangar capable of accommodating three Boeing 747s and two A320s side by side.

*Utilities*

The following utilities and services have been provided airport-wide

- potable and firefighting water
- HV and LV electricity supply and distribution
- stormwater drainage and related oil interception system
- sewerage system
- irrigation system
- gas distribution
- seawater for flushing and cooling
- security system

- voice and data communications
- aircraft approach and ground lighting.

Generally, all utilities have been placed within segregated reserves adjacent to primary road corridors (Fig. 12). This was a planning objective to ensure that future maintenance and upgrading of the utility systems would have minimal disruption to airport operations.

Development of the road and utility infrastructure is similar to a new town development, as in addition to the 34 million passengers who will pass through the airport in the opening year it is anticipated that some 45 000 people will be employed at the airport.

The provision of complete utility networks covering the entire airport represented a major design and construction effort. The complexity of the networks, the challenging construction sequence to suit third parties, the required early phasing of commissioning all contributed to the difficulties of effective site co-ordination. The use of 3-D computer models and clash analysis programmes assisted in this process.

*Roads*

The required capacity of the primary road access for travellers to the passenger terminal building has been greatly reduced by the implementation of the airport railway which forms the backbone for air passenger access. The principles adopted in the road network planning included segregation of private vehicles from commercial/industrial traffic; free-flow traffic management; minimal recirculation; provision of alternative emergency routes; and direct access to major land users.

Even with the estimated 43% of arriving and departing passengers taking the airport railway, it is estimated that by the year 2010 traffic flows on the main approach road to the passenger terminal building will exceed 3500 pcu (passenger car units) per hour in each direction. To the south, cargo and maintenance traffic will increase the flows to over 5000 pcu/h each way.

A dual three-lane expressway provides the trunk route for all traffic accessing the airport from the urban areas. On the airport the most southerly grade-separated interchange, which consists of five multi-span bridge structures, provides access to the cargo, catering and aircraft maintenance areas, effectively segregating the majority of heavy goods vehicles from passenger vehicles. The 1·5 km long primary access from this interchange to the passenger terminal building allows four lanes in each direction to facilitate effective weaving movements for traffic gaining access to the various levels and locations within the ground transportation centre.

It is vital to airport operations that the primary access corridor functions efficiently. As well as directional signage, dynamic flight information sig-

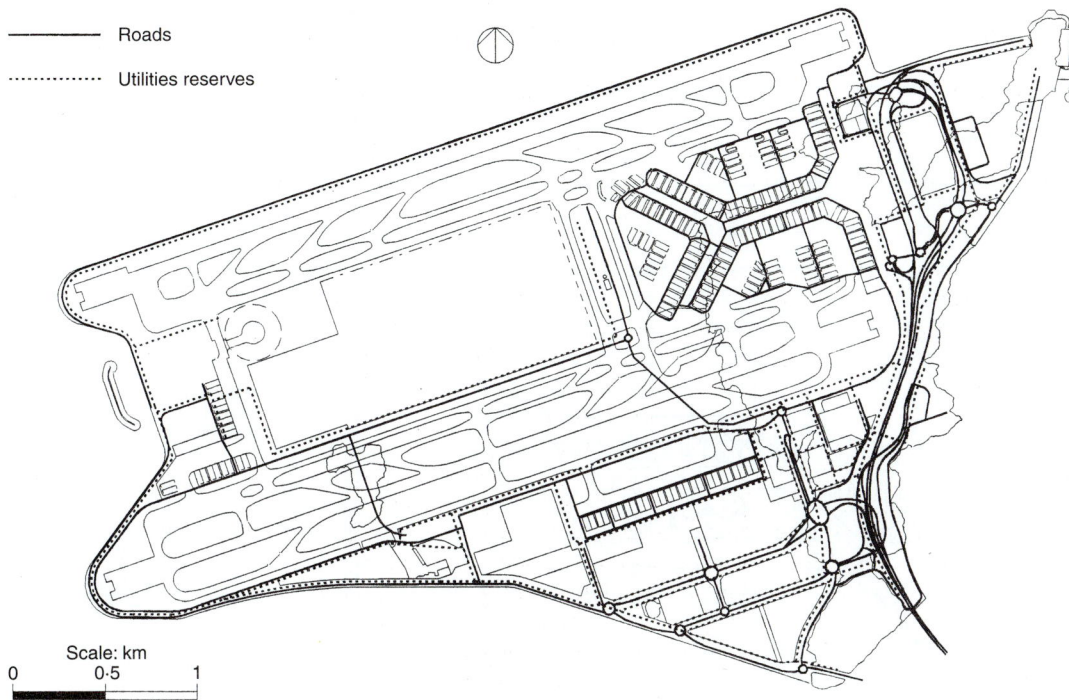

*Fig. 12. Road and
utilities network*

Roads

Utilities reserves

Scale: km

0    0·5    1

nage is also provided. The AA has implemented a comprehensive traffic control and surveillance system incorporating variable message signage to minimize disruption to passenger movements in the event of an incident on the approach roads.

*Ground transportation centre*

The focal point of the landside road network is the ground transportation centre located between the passenger terminal building and the proposed second terminal (Fig. 13), which facilitates passenger interchange between aircraft and the various ground transportation modes. These include the airport express railway, public buses and airbuses, tour coaches, hotel buses and limousines, baggage vans, taxis and private cars.

Segregated provisions have been made for each transport mode with emphasis placed on speed and efficiency of transfer, convenience, comfort and safety for all passengers, well-wishers and airport employees. At the core of the centre is the rail station[14] with direct air-conditioned links to the terminal. Departing air passengers travel along ramps with moving walkways to the check-in area within the terminal. Arriving air passengers have a similar ease of access from the terminal to their choice of ground transport, the rail departure platform being on the same level as air passenger arrivals (Fig. 14). The station is expandable to serve the future second terminal while maintaining the same level of service. Beyond 2010, up to 50% of air passengers and well-wishers are expected to enter and leave the airport through this station.

The ground level of the centre is occupied by the road arrivals traffic. Separate pick-up areas are provided for public buses, tour coaches, hotel

vehicles and taxis; the latter incorporates a 24-taxi simultaneous pick up system. Private car pick-up facilities for arriving passengers are provided in the multi-storey and at grade car parks.

The highest level of the centre is occupied by the road departures traffic. A departures forecourt and median strip provide a total kerb length of 650 m along the face of the processing terminal at airport opening. This allows taxis, public buses, hotel vehicles and private cars to drop off their passengers at this level.

*Passenger terminal building*

Whether travellers enter the check-in area of the passenger terminal building from the forecourt or from the station platform, they find themselves within a breathtaking interior space. The dominant feature is the roof, comprising 18 ha of steel barrel vaults (Fig. 15). The 36 m spans create a feeling of light and space and the geometry is aligned to lead passengers naturally through check-in, immigration and security towards their departure gate. The roof is high over the processing terminal, cantilevering 27 m over the forecourt, reducing in height over the east hall and along the central concourse before rising again over the west hall. Each 36 m x 36 m vaulted roof panel is formed of a lattice of straight 6 m long I-beams, 406 mm deep for the majority of the building and 457 mm deep members for the diagonal concourses. The architectural requirement for constant depth and flange width required varying forces within the members to be accommodated by changes in flange thickness. The roof geometry is revealed by exposing the bottom flange of all diagonal elements and within the triangles so formed are soffit lining panels (perforat-

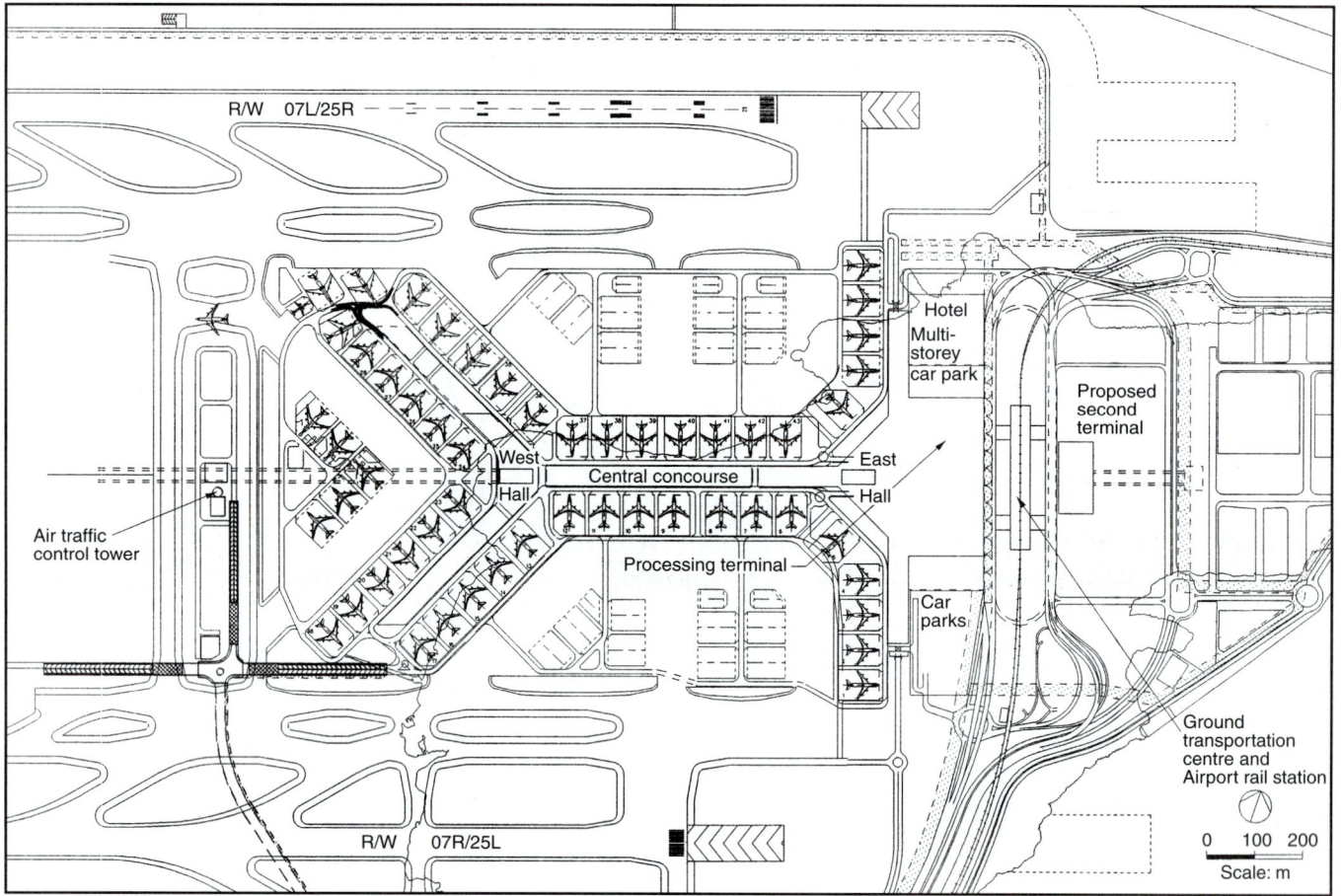

Fig. 13. (top). Plan of terminal building and ground transportation centre

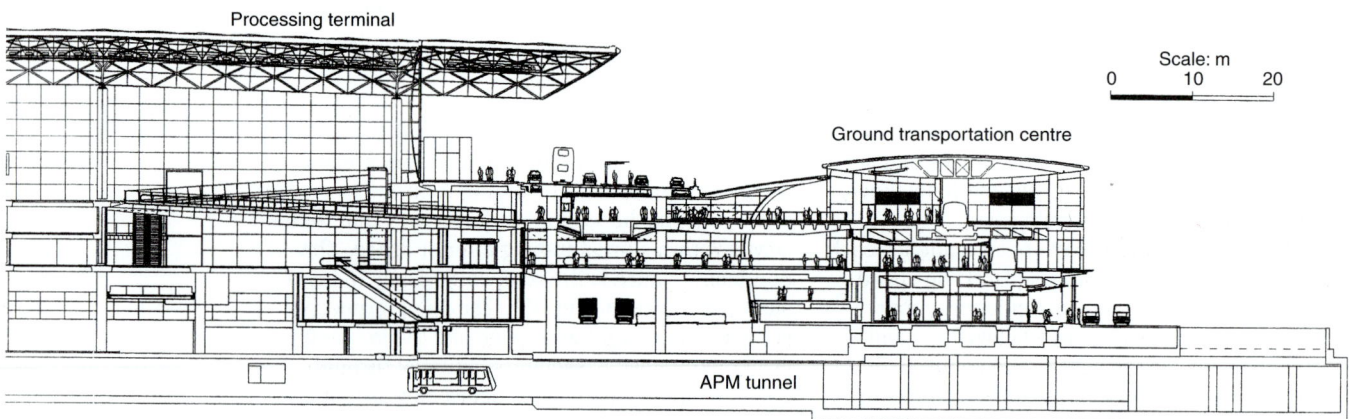

Fig. 14. (above). Section through processing hall and ground transportation centre

ed metal panels with acoustic insulation) positioned flush with the flanges.

Critical wind pressures were determined from wind tunnel tests which indicated that the Hong Kong code for wind loading was not conservative[15] for low-level structures in such exposed conditions. Seismic design was based on the New York code in the absence (at the time of design) of guidance for Hong Kong. The design required the impact of the sequence of roof module erection on the flexible concrete columns to be assessed to ensure that the final structure was within tolerance. Some vaults, primarily those over the processing terminal, are tied to reduce the impact of thrusts on the cantilever concrete columns.

Three important connection details emerged from the design. First, that of a typical node where six elements are joined: the solution adopted was a node plate as depicted in Fig. 16. The second was the column head connections (Fig. 17) where within the confines of a 1200 mm dia. reinforced concrete column top, two 100–150 t, 36 m span roof modules had to be located together with a 150 mm dia. downpipe and conduits for electrical services. The third was the connection detail to link the undulating roof with the

28

cladding: the solution was a cast stainless-steel armature connection or 'wishbone' (Fig. 18).

The roof covering comprises a waterproof membrane, a vapour barrier and thermal and sound insulation layers on a steel deck. Internally, the soffit lining panels reflect light from luminaries mounted on the roof gantry and from daylight which enters the building through the skylights.

The contract for the design of the passenger terminal building was awarded to the Mott Consortium in March 1992. The consortium comprised Mott Connell, the Hong Kong practice of Mott MacDonald (UK) and Connell Wagner (Australia), which was responsible for engineering design and consortium management, and Foster and Partners which undertook the architectural design, in association with BAA, which carried out operational planning and systems design. As subconsultant, Ove Arup and Partners designed the roof and architectural steelwork, W. T. Partnership was the quantity surveyor and O'Brien Kreitzberg provided design project scheduling.

The first phase of the building is designed to handle 35 million passengers per annum. The projected one-way standard busy rate (SBR: defined as the projected activity during the 30th busiest hour in the year; this is a more reliable measure of an airport's busy periods than a peak rate which will experience large fluctuations from year to year) for this annual throughput is 5500 passengers/h. The gross floor area of the building in the first phase is 515 000 $m^2$ and is over 1·3 km long, making it one of the largest single terminal buildings in the world. The concourses provide direct access through fixed link bridges and apron drive loading bridges to 38 aircraft stands, at airport opening, rising to 48 on completion of the north–west concourse (Fig. 13). An automated people mover and moving walkways will

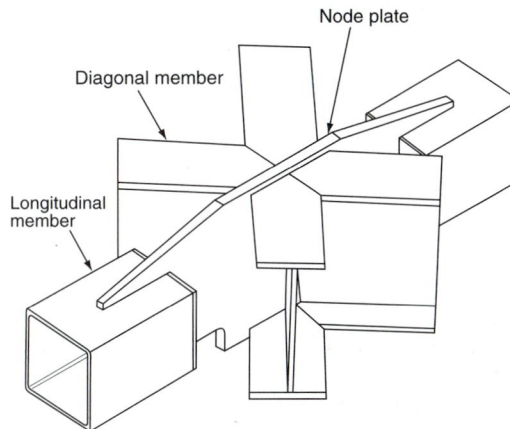

*Fig. 15. 3-D study internal view of roof*

*Fig. 16. Node plate detail*

ensure that people are able to move through the building quickly and comfortably.

Building services have been integrated within the structure to minimize obstructions to the passenger processing facilities and within the roof. Thus, from a double-height basement containing the chiller plant which draws in seawater for cooling, conditioned air is fed vertically to the various levels of the building by way of binnacles. Also, within the basement is the heart of the baggage handling system capable of handling 19 200 bags per hour at saturation. It incorporates a machine-automated explosive detection system which simplifies check-in as it enables all X-raying for hold baggage to be carried out within the baggage handling system.

Due to the large footprint of the building which straddles both the original island of Chek Lap Kok and reclaimed land, many different types of foundation have been employed including pad-and-strip footings, rafts, bored and H-piles. Below ground level, the basement and tunnels are designed as water-retaining structures and to

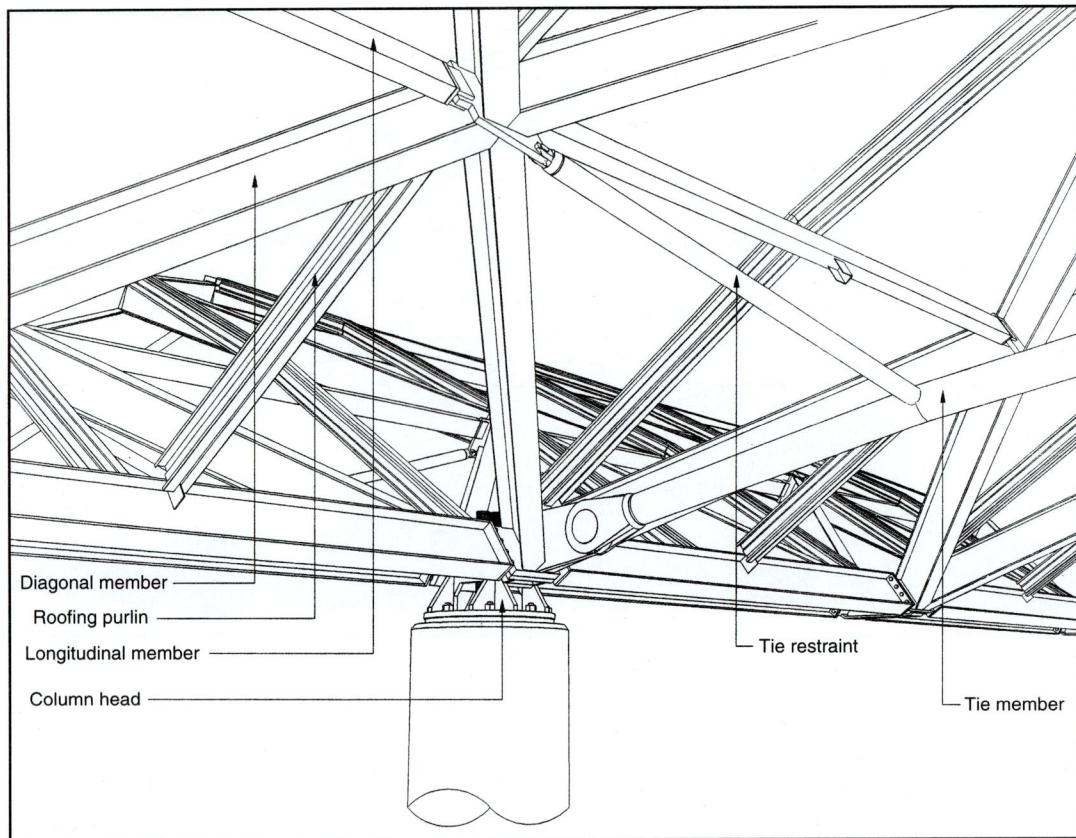

*Fig. 17. Column head
detail*

Diagonal member

Roofing purlin

Longitudinal member

Column head

Tie restraint

Tie member

*Fig. 18. (below).
Connection detail
between roof and
cladding*

counteract uplift pressures passive tension
anchors drilled into rock have generally been
used. This solution was determined to be more
economical in the long term than providing a
drained basement. A long section through the
processing terminal is shown in Fig. 14 and a
cross-section through the central concourse is
shown in Fig. 19. The automated people mover is
located in the central pair of tunnels and initially
operates along the central concourse between stations at the east and west halls. However, the tunnel has been constructed over 2 km long extend-
ing eastwards to the future processing terminal
and westwards beyond the cross-field taxiways in
readiness to serve the mid-field concourse development.

The main structural frame of the building is *in
situ* reinforced concrete. During the scheme
design six alternative forms of construction were
examined ranging from *in situ* concrete with various combinations of precast concrete to an all-
steel solution. The *in situ* concrete was selected
on cost and programme grounds. While the
troughed and coffered floor slabs were detailed to
offer the possibility of precasting, the contractor
chose an all *in situ* construction. Generally, the
structural frame comprises 700 mm deep beams
spanning 12 m with ribs at 1·5 m centres and a
125 mm thick slab between ribs. Two different
stability systems have been used. In the processing terminal horizontal loads are resisted by the
framing action of beams, ribs and columns. In the

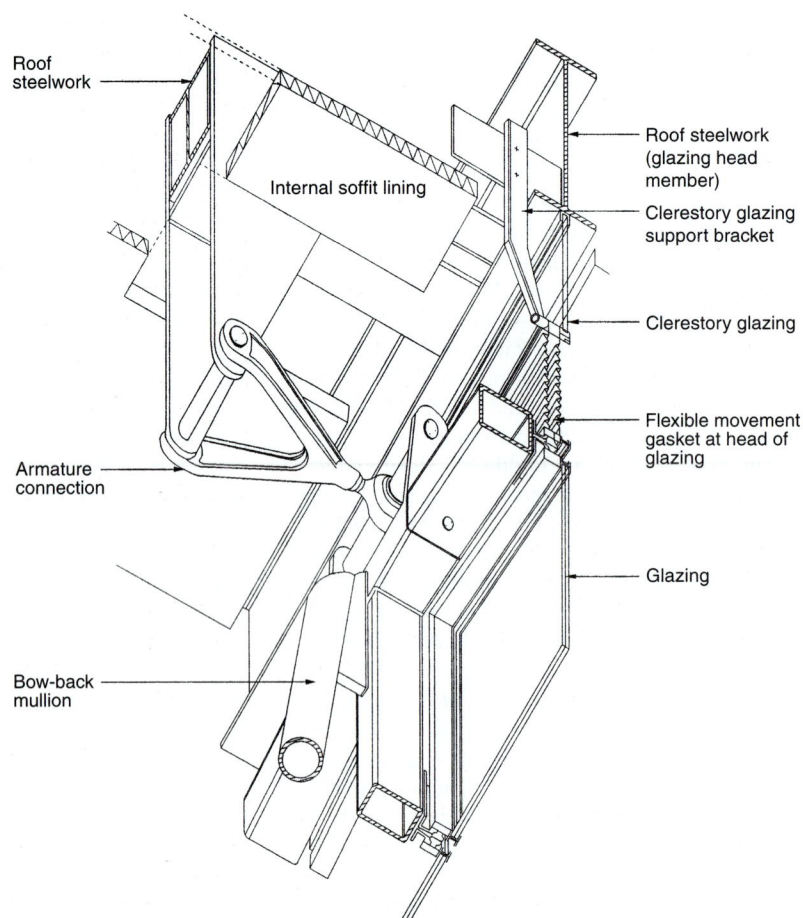

Roof
steelwork

Internal soffit lining

Roof steelwork
(glazing head
member)

Clerestory glazing
support bracket

Clerestory glazing

Armature
connection

Flexible movement
gasket at head of
glazing

Bow-back
mullion

Glazing

Skylight

Daylight reflector

Glazed
cladding

Structural steel cabins

Departure concourse

FFL + 16·150

Arrival concourse

Office

Office

FFL + 12·150

Metal cladding

Ramp    accommodation

FFL + 7·600

APM  tunnel

Tunnel for
future
baggage
system

Scale: m
0    2·5    5

00·000 datum

Services tunnel            Services tunnel

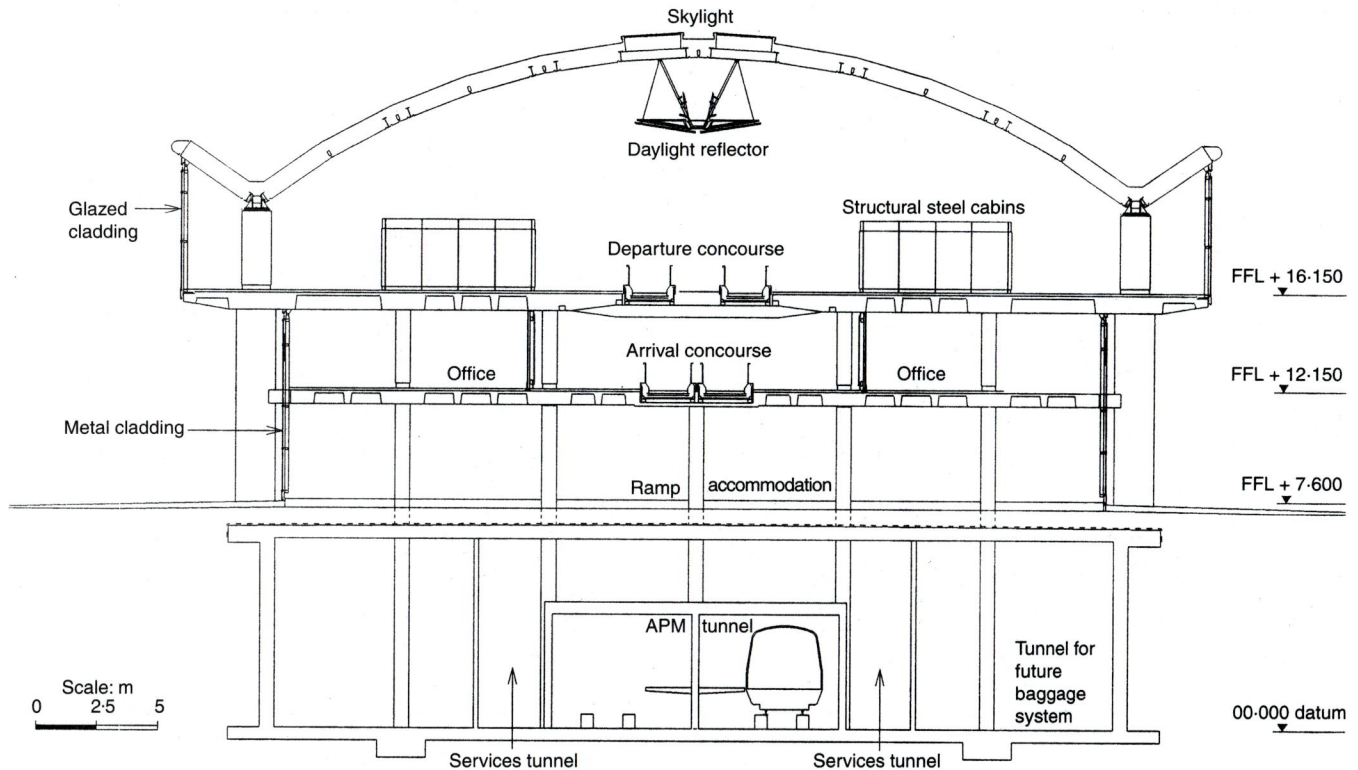

concourses lateral stability is provided by shear walls which have been carefully positioned to avoid constraining flexibility to reconfigure the accommodation. Within the main concrete frame, retail, office and toilet cabins are steel framed as are the fixed-link bridges linking the concourses with the aircraft loading bridges.

The departures level cladding comprises a pan-elized glass system with aluminum frames supported by bow back steel mullions (Fig. 20). The glazed wall is up to 28 m high allowing daylight to penetrate the open plan terminal. Laminated toughened glass with a polyvinyl butyral interlay-er and internal coating has been used to achieve the necessary solar control. The cladding below departures level is made up of a combination of glazed, solid steel and louvered panels which are modularized to allow for future changes to requirements, which is a common event in ramp-level accommodation.

The building has been designed to facilitate future expansion and at present the design and construction of a 35 000 m² extension to the north-west concourse is underway. Provision has also been made on the north and south sides of the processing terminal to extend the building by 72 m (two roof bays) on each side.

## Environmental initiatives

Right from the planning stage the AA has been committed to making the new airport one of the most environmentally acceptable infrastructure projects in Hong Kong, and indeed the region. A major reason for the choice of Chek Lap Kok for the new airport was on environmental grounds as

more than 350 000 people are directly affected by the major flight paths of Hong Kong's Kai Tak Airport; whereas, it is estimated that fewer than 100 people will experience high noise levels at the new airport where the aircraft approach and departure tracks are over the sea. Fig. 21 com-pares the more stringent 25 NEF (noise exposure forecast) contour for the new airport at the year 2000 with the 30 NEF contour at Kai Tak Airport.

An environmental impact assessment was undertaken as part of the NAMP to investigate the consequences of construction and operation of the airport and to recommend mitigation mea-sures where necessary. Because of uncertainties

*Fig. 19. (top). Cross-section through central concourse (elevations in mm)*

*Fig. 20. (above). Architectural perspective view of meeters and greeters cladding*

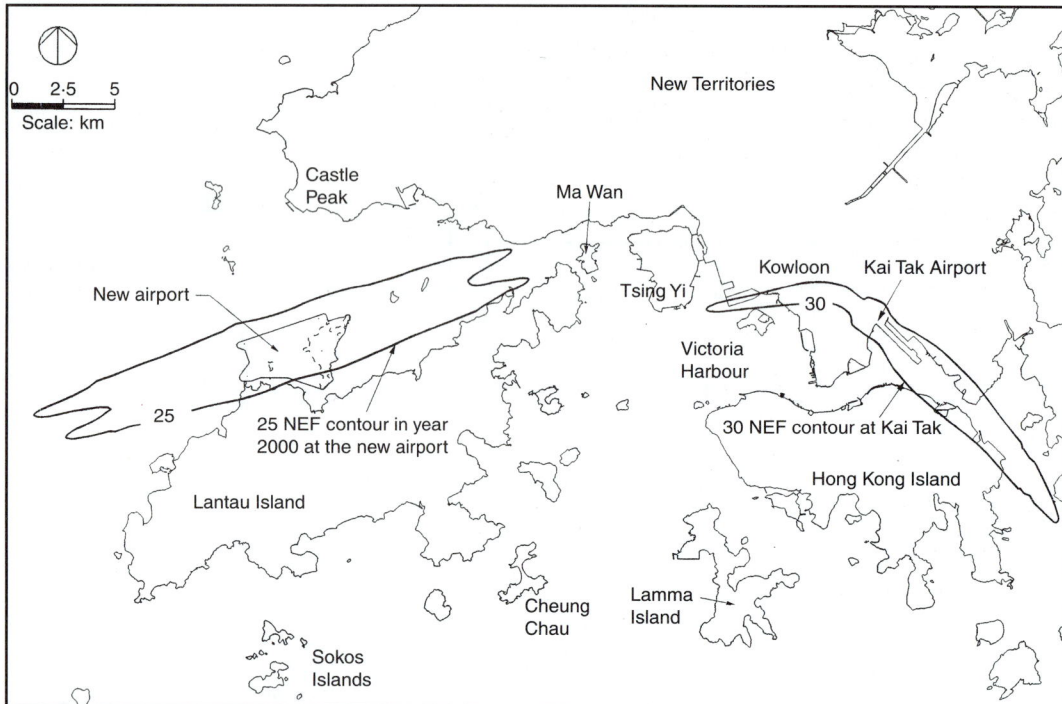

*Fig. 21. Comparison of noise exposure forecast contours*

*Fig. 22. (below). Bubble curtain to alternate sound waves*

in projected construction and operation methods a conservative approach was taken and during the detailed design environmental reviews were updated and designs and mitigation measures modified. Throughout the design and construction stages the AA reported regularly and sought agreement from Hong Kong's various environmental agencies. The AA established its own environmental group within its engineering department to monitor environmental issues and plan for operations. To achieve its environmental aims the AA incorporated a number of practical measures into the design, such as a 'grey water' treatment plant which recycles waste water from the passenger terminal building, catering facilities and aircraft wash stands for irrigation purposes, interceptors built into the drainage system in aircraft fuelling areas to stop potential pollutants from flowing into the sea and fuel spill containment traps at stormwater outfalls. Dedicated systems have also been designed for the treatment of effluents arising from the fire training facility as well as the AA's vehicle wash facility.

Water quality of the sea around the airport has been given a high priority and is regularly monitored. Particular attention has been paid to potential effects on the Chinese white dolphins (*Sousa chinensis*) found in these waters. An aviation fuel transfer facility has been constructed close to the airport and with the advice of cetacean experts the AA has implemented, in co-operation with the aviation fuel operator and its contractor, all practicable environmental mitigation measures which have been recommended. During construction, these measures included the installation of an underwater 'bubble curtain' to attenuate sound pressure

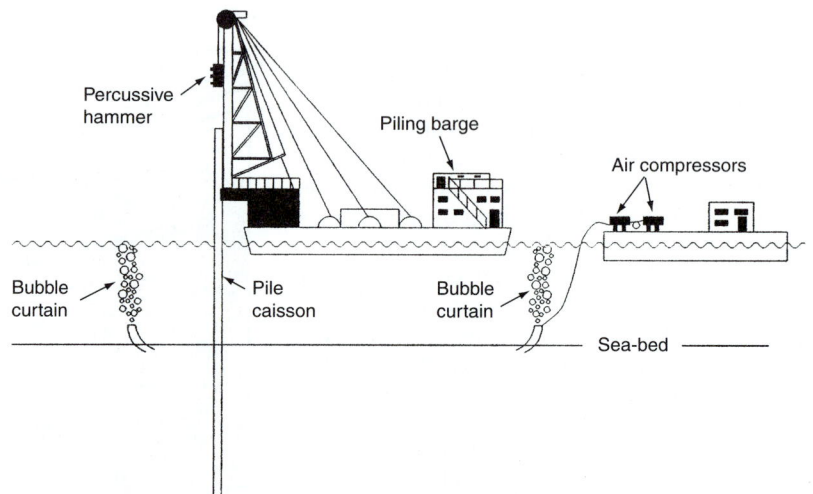

waves during piling operations (Fig. 22).

Key to the airport operating framework is the AA's requirement for its business partners to prepare and implement environmental management plans for their operations. These plans require each operator to identify key issues related to air or noise emissions and effluent discharges, detail licensing points and present an environmental management strategy. The airport community has co-operated well in the preparation of these plans. Experience gained from the overall process is that detailed environmental impact assessments must be incorporated at the earliest possible stages of project planning and be continued throughout the design phase, targeting stakeholders as soon as possible. The drafting, monitoring, audit and mitigation processes are major responsibilities for all airport organizations.

## Concluding remarks

Hong Kong is used to accepting the challenges presented by major engineering projects. This brief paper has attempted to summarize the key planning and design tasks involved with the new airport which has dwarfed previous Hong Kong efforts and perhaps ranks as one of the most significant of engineering projects anywhere in recent times.

To achieve the objectives within a time scale of less than eight years is a tribute to the massive human endeavour which has had to be undertaken.

## Acknowledgments

The authors wish to thank the AA and the government of the Hong Kong Special Administrative Region for permission to publish this paper. They also acknowledge with appreciation the enormous contribution made by design consultants, contractors and staff of the AA in the planning, design and construction of the new airport. The authors would also like to thank in particular C. C. Calton, C. Leeks, R. I. Livingston, L. Mujaj and Dr A. R. Pickles for their assistance in the preparation of this paper.

## References

1. PLANT G.W., COVIL C. S. and HUGHES R. A. (eds) *Site Preparation for the New Hong Kong International Airport.* Thomas Telford, London, 1998.
2. RMP-ENCON. *Hong Kong Air Transport System Long Term Planning Investigation Executive Summary*, March 1975.
3. RMP-ENCON. *Replacement Airport at Chek Lap Kok—Master Plan Consultancy Final Report*, November 1982.
4. GREINER-MAUNSELL. *New Airport Master Plan Final Report*, December 1991.
5. OAKERVEE D. E. The planning, design and construction parameters of Hong Kong's new airport. *Proceedings of the Institution of Civil Engineers, Transport*, 1994, **105**, Nov., 235–247.
6. AIRPORT AA OF HONG KONG. *Annual Report 1996/1997.*
7. THOMSON A. and OAKERVEE D. E. The construction of Hong Kong's new airport. *Proceedings of the Institution of Civil Engineers, Civil Engineering, Hong Kong International Airport, Part 1: airport*, **126**, *1998, 35-54.*
8. PICKLES A. R. and PLANT G. W. An overview of the geotechnical aspects of the airport project. *HKIE Geotechnical Annual Seminar*, 29 May 1998, Hong Kong.
9. RMP-ENCON. *Replacement Airport at Chek Lap Kok—Civil Engineering Studies. Report No. 2 , Test Embankment.* December 1982.
10. UK PSA. *A Guide to Airfield Pavement Design and Evaluation*, 1989.
11. MUJAJ L. and PLANT G.W. The design philosophy for the airfield pavements for Hong Kong's new airport. *Third International Conference on Road and Pavement Technology*, 28–30 April 1998, Beijing.
12. US FAA. *Airport Pavement Design for the Boeing 777 Airplane.* Advisory Circular 150/5320-16, 1995.
13. MUJAJ, L. Concrete block paving at the new airport for Hong Kong Third international workshop on Concrete Block paving, Cartagena de Indias, Columbia, 10-13 May, 1998.
14. Design and construction of the airport railway. *Proceedings of the Institution of Civil Engineers, Civil Engineering, Hong Kong International Airport, Part 2: Road and Rail Links*, **126**, 1998.
15. Scott, D M et al. The design of the passenger terminal roof at Chek Lap Kok. In preparation.

Appendix 1. Hong Kong's airport facilities—comparative statistics

| | Existing airport at Kai Tak | New airport at Chek Lap Kok Phase 1(a) |
|---|---|---|
| **Total airport site** | 333·8 ha | 1 248 ha |
| **Annual passengers** | | |
| actual passengers in 1996 (excl. transit) | 29·5 million | |
| design capacity | | 35 million |
| **Air cargo tonnage** | | |
| actual cargo in 1996 | 1·56 million | |
| design capacity | | 3 million |
| **Airfield** | | |
| Runways | 1 | 1 (2)* |
| Runway length | 3393 m | 3800 m |
| Taxiway system | 7·1 km | 26 km (35 km)* |
| Apron area | 1·03 km² | 1·25 km² (1·33 km²)* |
| **Total passenger terminal area** | 66 000 m² | 515 000 m² (550 000 m²)* |
| retail area | 13 600 m² | 30 000 m² (31 300 m²)* |
| retail outlets | 40 | 120 |
| **Aircraft gates** | 69 | 78 (88)* |
| frontal (connected to terminal by airbridge) (48)* | 8 | 38** |
| remote | 56 | 27 |
| air cargo | 5 | 13 |
| **Check-in counters** | 210 | 288 |
| **Immigration control desks** | 170 | 224 |
| arrivals | 92 | 128 |
| departures | 78 | 96 |
| **Customs inspection positions (arrivals)** | 52 | 76 |
| **Security screening positions (departures)** | 12 | 16 |
| **Baggage reclaim units** | 6 | 12 |
| **Passenger facilities** | | |
| moving walkways | 0 | 48 |
| automated people mover | 0 | 1 |
| car parking spaces | 1732 | 3100 |
| taxi loading spaces | 15 | 24 |
| bus stops | 6 | 17 |
| lounge seating | 3812 | 12 530 |
| flight information display boards | 283 | 2000 |

**Note:**
* Following commissioning of second runway, extension of the north-west concourse and associated facilities.
** Temporary closure of one frontal gate is planned due to the construction work for the extension of the north-west concourse.

| Contract Number | | Consultant/contractor | Date of award |
|---|---|---|---|
| 100 | Masterplan design | Greiner-Maunsell | July 1990 |
| 101 | Terminal building detailed design | The Mott Consortium | Mar. 1992 |
| 105 | Site preparation contract design | Greiner-Maunsell | Aug. 1992 |
| 106 | Lok On Pai refurbishment design | Llewellyn-Davies | Sept. 1992 |
| 110 | Airport expressway, rail link and roads design | Halcrow Asia Partnership Ltd | Sept. 1993 |
| 111A | Ground transportation centre and approach roads design | Ove Arup and Partners (HK) Ltd | Oct. 1993 |
| 111B | Airport railway design (GTC) | Ove Arup and Partners (HK) Ltd | Oct. 1993 |
| 120 | Temporary utilities design | Hyder Consulting Ltd (formerly Acer Consultants (Far East) Ltd) | May 1993 |
| 121 | Permanent utilities design | Parsons Brinkerhoff (Asia) Ltd | May 1993 |
| 122 | Stormwater, sewerage and irrigation design | Scott Wilson Kirkpatrick (HK) Ltd | May 1993 |
| 125 | Fuel distribution system design | Parsons Brinkerhoff (Asia) Ltd | Oct. 1993 |
| 126 | Aviation fuel receiving facility design | MMBP-Atkins | Oct. 1994 |
| 127 | Feasibility study and environmental impact assessment for aviation fuel pipeline | Montgomery Watson HK Ltd | June 1995 |
| 140 | Software quality management | Engineering Centre for Software | Dec. 1995 |
| 141 | Communications systems design | Parsons Brinkerhoff (Asia) Ltd | Oct. 1993 |
| 142 | Security systems design | Parsons Brinkerhoff (Asia) Ltd | Oct. 1993 |
| 145 | Waste management study | Montgomery Watson HK Ltd | Sept. 1993 |
| 150 | Airfield tunnels design | Hyder Consulting Ltd (formerly Acer Consultants (Far East) Ltd) | May 1993 |
| 151 | Airfield pavements design | MMBP-Atkins | July 1993 |
| 152 | Airfield ground lighting design | Gibb-Harris and Sutherland | July 1993 |
| 153 | Apron lighting design | Rust Asia Pacific Ltd | July 1993 |
| 155 | Perimeter fence and ancillaries design | Mouchel Asia Ltd | Nov. 1993 |
| 156 | Northern runway and north-west apron design | MMBP-Atkins | July 1996 |
| 157 | Northern runway and north-west apron airfield lighting design | Gibb-Harris and Sutherland | July 1996 |
| 162 | Airport ancillary buildings design | Ho and Partners | Jun 1995 |
| 170 | Landscaping design | Scott Wilson Kirkpatrick (HK) Ltd | Apr. 1994 |
| 190 | New airport transport study | Wilbur Smith Associates | July 1992 |
| 192 | Noise mitigation measures study | Llewellyn-Davies | Oct. 1992 |
| 193 | Urban design guidelines and study | Urbis Travers Morgan Ltd | Dec. 1993 |
| 199G | Marine geology of Chek-lap Kok | British Geological Survey | Aug. 1993 |
| 202 | Supply of geotechnical instruments | Sinco Slope Indicator Co. (Furgo) | Jan. 1993 |
| | | RS Technical Instruments | Jan. 1993 |
| | | Geokon Inc. | Jan. 1993 |
| | | Geotechnical Instruments (HK) Ltd | Jan. 1993 |
| 255 | Site investigation and laboratory testing (land) | Intrusion-Prepakt Foundation Ltd | Sept. 1993 |
| 256 | Marine site investigation and laboratory testing | Lam Geotechnical Ltd | Oct. 1993 |

Appendix 2.
Consultants and contractors concerned with design of Hong Kong's new airport

# Hong Kong International Airport— construction

*A. I. Thomson, BSc, CEng, MICE, and D. E. Oakervee, CEng, FICE*

The new Hong Kong International Airport was built on a remote island site at Chek Lap Kok in just seven and a half years. This paper describes how construction progressed from the initial advance works contract awarded in January 1991 to the opening of the completed, state-of-the-art airport and its massive 515 000 m² terminal building in July 1998. It details the HK$ 44 billion package of contracts let by the main client, the Hong Kong Airport Authority, and the methods of construction management used to ensure safe completion both on time and within budget. It also addresses the unique construction support facilities and services which were provided by the Authority to support the 21 000 people working on site during the peak construction phase.

*Proc. Instn Civ. Engrs,*
*Civ. Engng, Hong Kong*
*International Airport ,*
*Part 1: airport,* 1998,
**126,** 35-54

*Paper 11523*

*Written discussion closes*
*15 November 1998*

On 29 November 1995 the airport committee of the Sino-British joint liaison group signed agreed minutes on membership of the Hong Kong Airport Authority (AA), which paved the way for establishment of the AA on 1 December 1995. Prior to this, the planning, design and construction had been progressing on a phased basis under the auspices of the provisional airport authority (PAA) to complete the airport to the maximum extent possible by June 1997.

Any civil engineering project evolves on a logical progressive basis. The airport project evolved over the years by taking account of, not only engineering logic, but also the incremental release of funds to the PAA as the project progressed. This step-by-step funding had a significant influence on how the construction strategy developed. This paper follows the chronology of construction contract awards and sequence of construction.

The airport is one of the ten projects that together make up the government's Airport Core Programme as described elsewhere in a companion paper.[1] The factors which influenced the design of the various airport facilities are also described in a companion paper.[2]

A list of all significant contractors employed in the construction of the airport project is given in Appendix 1.

Fig. 1. *Project construction programme*

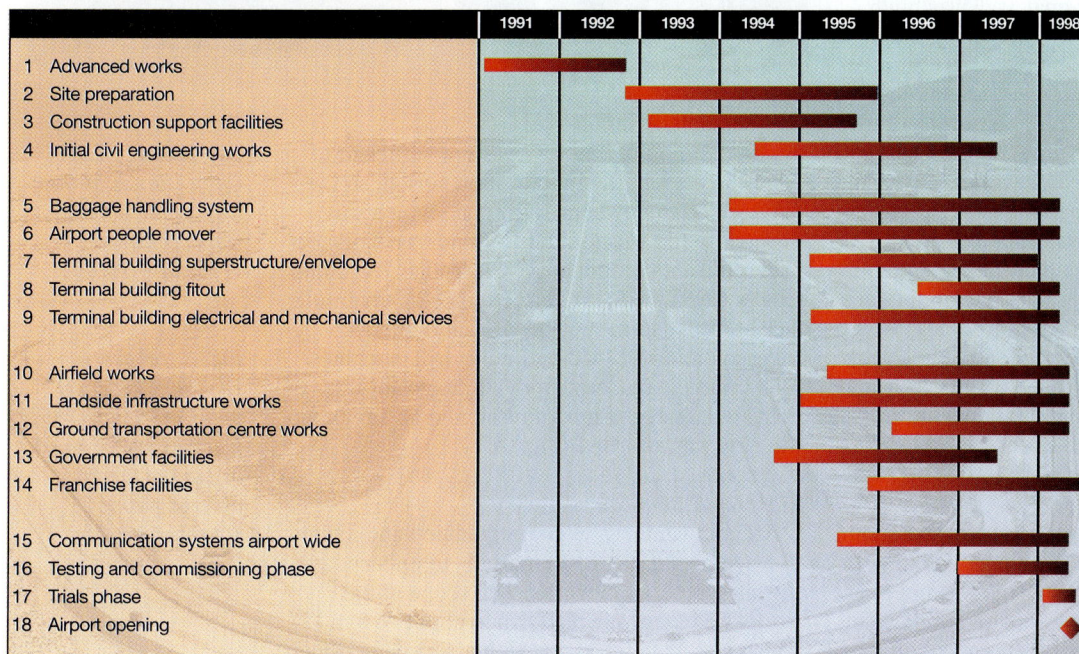

|  | 1991 | 1992 | 1993 | 1994 | 1995 | 1996 | 1997 | 1998 |
|---|---|---|---|---|---|---|---|---|
| 1 Advanced works | | | | | | | | |
| 2 Site preparation | | | | | | | | |
| 3 Construction support facilities | | | | | | | | |
| 4 Initial civil engineering works | | | | | | | | |
| 5 Baggage handling system | | | | | | | | |
| 6 Airport people mover | | | | | | | | |
| 7 Terminal building superstructure/envelope | | | | | | | | |
| 8 Terminal building fitout | | | | | | | | |
| 9 Terminal building electrical and mechanical services | | | | | | | | |
| 10 Airfield works | | | | | | | | |
| 11 Landside infrastructure works | | | | | | | | |
| 12 Ground transportation centre works | | | | | | | | |
| 13 Government facilities | | | | | | | | |
| 14 Franchise facilities | | | | | | | | |
| 15 Communication systems airport wide | | | | | | | | |
| 16 Testing and commissioning phase | | | | | | | | |
| 17 Trials phase | | | | | | | | |
| 18 Airport opening | | | | | | | | |

*Alistair Thomson is head of construction at the Hong Kong Airport Authority*

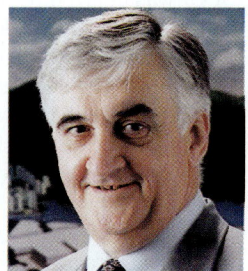

*Douglas Oakervee is project director at the Hong Kong Airport Authority*

## Budget and programme

Construction of the airport was initially based upon the recommendations of the *New Airport Master Plan*.[3] This envisaged a construction programme starting at the beginning of 1992 with a target date for opening of mid-1997, together with a costed implementation programme which set the project budget control total. It was not possible to implement fully these plans until the PAA was reconstituted as the AA in December 1995. In 1996 the government announced the new airport would open in April 1998, and at that time the construction programme was revised and the costed implementation programme was updated. The construction programme (Fig. 1) was implemented in five phases

- advance works
- site preparation
- initial contracts
- main works
- integration and trials.

## Advance works

### Advance works contract (AWC)

In 1991 the project site comprised two islands. The larger one, Chek Lap Kok (302 ha), was inhabited by three fishing villages and the smaller one, Lam Chau (8 ha), was uninhabited. Following rehousing of the villagers, an advance works contract was awarded in January 1991. The purpose of this contract was to provide basic infrastructure on the otherwise barren island of Chek Lap Kok. This was done by removing the tops of the three highest hills from +120 m PD (project datum) to form plateaux at about +95 m PD. The material arising from this was used to reclaim a bay to + 6 m PD, which is about 4·7 m above mean sea level, in the north-east corner (Fig. 2). Here a heavy-duty quay wall was built together with a pair of 1 MW electricity generators, a potable water treatment plant, raw and sea-water distribution systems, offices, accommodation blocks and a canteen. This was sufficient to provide facilities to support the AA's own staff and the site preparation contractor's workforce during the intensive mobilization period.

The award of the site preparation contract (SPC) was delayed by seven months due to funding difficulties but work was able to continue on a step-by-step basis by utilizing the dredging resources of the advance works contractor. A completion certificate was issued for the AWC on the same day that the site preparation contract was awarded on 30 November 1992.

### Pre-ordered earthworks plant

It was also recognized that the site preparation contractor would require a significant amount of large-scale earth-moving plant. This was not readily available on the world market and purchase of such equipment would have taken up to 6 months

to manufacture and deliver.

The duration of the SPC was not sufficient to allow such a lengthy mobilization period, so the AA decided to buy some new plant which was novated to the successful SPC tenderer. This pre-ordered plant comprised 17 Caterpillar 785 dump trucks (136 t payload), four Demag 285 hydraulic face shovels (19 m³ bucket capacity) and six Ingersoll-Rand DMM2 hydraulic drill machines (250 mm hole dia. capacity). All this plant was delivered, tested and commissioned on site before the SPC was awarded (Fig. 3).

### Site preparation contract (SPC)

Six joint ventures were prequalified and invited in November 1991 to submit bids to undertake the site preparation works. The scope of the works was to form a platform at about +6·5 m PD, which was 1248 ha in size with a perimeter about

*Fig. 2. (top). Chek Lap Kok and Lam Chau islands at the time of the advance works contract (February 1992).*

*Fig. 3. (above). Demag 285 shovel and Caterpillar 785 truck assembly on site as part of the pre-ordered plant (October 1992)*

17 km long. The works would involve dredging about 60 Mm³ of unsuitable marine mud, removing about 95 Mm³ from the island of Chek Lap Kok, Lam Chau and the two Brothers Islands and placing the material in reclamation, and importing and placing about 86 Mm³ of marine sand. All work had to be completed within 31 months of the award of contract.

When tenders were returned in March 1992 they were significantly above the AA's budget and parallel tender estimate. A way had to be found for the AA to become less dependent on the world's limited resource of trailer dredgers. This was done by inviting tenderers to submit alternative tenders based on the use of a land borrow area (instead of importing marine fill) and increasing the contract period to 41 months. The alternative tenders were submitted in June 1992 and were within budget and found to be competitive. The PAA did not have funds available to award the contract but a letter of intent was sent to the lowest and conforming alternative tenderer in July 1992. The contract sum was HK$ 9·1 billion. During the interregnum the AA negotiated extensions to the scope of works with the advance works contractor so that critical work to remove marine muds could continue.

The joint-venture contractor was able to mobilize quickly using the facilities provided under the AWC and the pre-ordered earthworks plant. The joint-venture also had fully developed its sub-contracting strategy during the tender period. The contractor was therefore in a position to award major sub-contracts quickly for the dredging works, land excavation works, seawall construction and geotechnical instrumentation works. Large earth-moving plant of the scale demanded by the SPC had never been used in Hong Kong and it was necessary to bring skilled workers from overseas. The Hong Kong government had introduced a special labour importation scheme whereby contractors could bring workers from overseas for fixed-term contracts to assist the construction of the Airport Core Programme projects. The first of these arrived in February 1993 to coincide with the start of production work on site.

The first task was to remove the unsuitable marine muds from the sea-bed to expose alluvial material which had a net cone penetration test tip resistance of more than 500 kPa. The depth of water varies from zero, adjacent to Chek Lap Kok and Lam Chau, to 15 m in places. The average thickness of mud which had to be removed was 8 m, increasing to 15 m in places. Most of the mud had to be taken to the government marine dump south of Cheung Chau some 40 km away from the site. Grab dredgers had to be used in shallow waters. More economical trailer suction hopper dredgers (Fig. 4) were used in waters where the depth exceeded 5 m. At peak production 10 trailer dredgers were used for mud removal. These varied in hopper capacity from 2500 m³ to 14 000 m³.

Table 1. Typical Chek Lap Kok drill-and-blast parameters

| Drilling-and-blasting parameters (primary) | Quantity | Units |
|---|---|---|
| Chek Lap Kok drill quantity | 65·3 | Mm³ |
| Annual drill requirement | 24·0 | Mm³/year |
| Hole diameter | 216 | mm |
| Bench height | 15 | m |
| Typical drill pattern | 6·5 x 7 | m x m |
| Drill penetration rate | 23 | m/h |
| Drill productivity | 850 | m³/h |

Fig. 4. The J.F.J. de Nul trailer suction hopper dredger working at Chek Lap Kok in 1994—the largest capacity dredger in the world at the time.

During a 36 month period a total of 70 Mm³ of mud were removed from site, with the maximum output of 14 Mm³ in one month.

The land-based earthworks was sub-contracted into two approximately equal parts. The sub-contractor in the northern half adopted a philosophy of using a mixed fleet of new and old trucks of varying capacity but with a restricted speed limit. The sub-contractor in the southern part used a single size fleet of new trucks operating at a higher speed limit. Therefore, the two sub-contractor's operations were physically segregated to avoid safety problems that would have been associated by mixing the two philosophies. Consequently, the safety record was good although there was regrettably one fatal accident involving a haul truck.

The joint venture sub-contracted the supply of explosives. This sub-contractor provided a down-the-hole service to the two earthworks sub-contractors. Packaged explosives and detonators were issued from a site-based magazine under the supervision of the mines division of the Hong Kong government. Ammonium nitrate and fuel oil (ANFO) was manufactured in an emulsion plant on site and delivered to the face by 12 t capacity mix-and-place trucks. The geology of Chek Lap Kok is highly variable and therefore each blast had to be uniquely designed. Where 13 m benches were formed drill holes would be 216 mm dia. with 2·0 m sub-drill, 9 m of explosive column and

Fig. 5. Sham Wan village—one of three accommodation camps constructed to house the workforce living on site and Chek Lap Kok (May 1996)

6 m of stemmings produced from drillings. Typical blast parameters are given in Table 1. Explosives weighing 25 t would yield about 50 000 m³ of product and explosive consumption was 80 t/day on average. A daily feature was the shot firers' meeting when all parties would send a representative to be briefed about the blasts that had been designed and planned for the following day. There were two blasting periods; one at lunch time and the other being at the shift change at the end of the day before dusk. The blast evacuation zone was generally 500 m. There was not one injury from a blasting accident throughout the whole contract.

The product was categorized at the face into one of three gradings of material. Type A material was rock of particle size of 2 m or less. This material was taken and end-tipped on the alignment of the southern runway zone. Type B material was decomposed rock of which 100% was less than 300 mm in size. This material was end-tipped in areas at which it was anticipated that buildings would have to be constructed on piled founda-

tions. In practice it was found that much material was generated which did not fall into one or other of these specified gradings. It was inevitable that a mixture of these two sorts would arise due to the variable geology. This material was classified as type A/B and a pragmatic approach adopted whereby the contractor was allowed to place unprocessed type A/B mixtures in those areas where it was not essential to use either type A or type B. The bulk of material was end-tipped into the sea but in areas where the factor of safety of the advance face was less than 1·1 it was necessary to place a blanket of material in advance by using bottom-dump barges. Soundings were taken in front of the advancing tip faces to ensure that mud waves were not building up.

Seawalls were constructed in parallel to reclamation works. Dredging to a specified minimum level or minimum CPT value of 680 was carried out using grab dredgers which removed a total of 9 Mm³ of mud. A total of 2·1 km of vertical seawalls were made from concrete blocks and 12 km of sloping seawalls constructed using rock products arising from Chek Lap Kok. Filter layers and armour rock was processed by taking product from the blasted face and sorting using a combination of grizzly screens, back hoes and grapples. The sorted products were then loaded on to barges and towed to the site and placed by grab barges.

The contractor's alternative tender was based on the importation of marine sand to provide the balancing quantity of material required to complete the platform. This was done using four borrow areas at the Brothers Islands, east of Sha Chau, Urmston Road and Ninepins. A total of 46 Mm³ of overburden mud had to be removed so as to expose suitable filling material. A fleet of up to 18 trailer dredgers (Table 2) worked continuously for 20 months delivering material 24 hours a day, 7 days a week. This was known as the 'sand train' and it delivered a total of 86 Mm³, some of which was dumped in its final position, some was re-handled by cutter dredger and some was pumped ashore.

When dealing with such very large quantities of material it is essential to monitor progress rigorously to ensure programme objectives are being achieved. This was done using state-of-the-art computer survey techniques. Hydrographic surveys were carried out utilizing the AA's own dedicated survey launch operating independently of the contractor's own survey units. Land excavation and reclamation was surveyed by aerial photographs which were scanned to produce orthophotos and digital terrain models from which volumes and areas were readily calculated.[4]

The final account was settled for a sum within budget some 37 months after award of contract which was four months before the planned contract completion date. This success was due to a number of factors where the AA and the contractor worked together, such as

- high technical and managerial skills
- overall quantities were not varied from those initially contracted
- continuity of work for the dredgers and earthworks plant
- pragmatic utilization of land-based materials 'as dug'
- quick and accurate volume surveys to monitor progress.

## Construction support
The airport at Chek Lap Kok had characteristics that were unique in the history of construction industry of Hong Kong, some of which are

- it was an offshore site
- it was remote from the urban areas
- no infrastructure or municipal services were available
- a workforce in excess of 21 000 existed
- overfull employment in construction industry was predicted

Table 2. Trailing-suction hopper dredgers used at Chek Lap Kok

| Trailer | Hopper capacity: m³ | Mud: Mm³ | Sand: Mm³ | Total: Mm³ |
|---|---|---|---|---|
| J.F.J. de Nul | 11 750 | 7·1 | 8·1 | 15·2 |
| Lelystad | 10 330 | 4·0 | 6·2 | 10·2 |
| Vasco Da Gama | 10 156 | 7·8 | 0·2 | 8·0 |
| Volvox Hollandia | 9000 | 5·1 | 9·8 | 14·9 |
| Prins der Nederlanden | 8960 | 11·2 | 0·5 | 11·7 |
| Geopotes 15 | 8825 | 2·3 | 3·2 | 5·5 |
| Geopotes 10 | 8645 | 7·7 | 18·1 | 25·8 |
| HAM 310 | 8225 | 4·4 | 14·2 | 18·6 |
| Delta Queen | 8128 | 4·1 | 6·1 | 10·2 |
| Cornelis Zanen | 8000 | 5·3 | 3·3 | 8·6 |
| Barnet Zanen | 8000 | 1·9 | 7·1 | 9·0 |
| Volvox Delta | 8000 | minor | quantities | |
| Volvox Hansa | 6000 | 8·8 | 0·1 | 8·9 |
| Rigelstar | 6000 | 3·9 | 9·0 | 12·9 |
| Sanderus | 5000 | 6·5 | 0·0 | 6·5 |
| Apollo | 4850 | 9·4 | 0·0 | 9·4 |
| Johanna Jacoba | 4100 | 1·1 | 0·0 | 1·1 |
| Coronaut | 2500 | 7·1 | 0·0 | 7·1 |
| **Total** | | **97·7** | **85·9** | **183·6** |

Table 3. Construction support facilities

| Facility | Quantity |
|---|---|
| Vertical seawall | 2·1 km |
| Marine cargo handling areas | 5 |
| Roll-on/roll-off berth | 1 |
| Ferry pier for 1500-seater boats | 1 |
| Small boat harbour | 1 |
| Pontoon landing points | 4 |
| Transshipment centre on the mainland | 1 |
| Electricity supply at 11 kV | 21 kVA |
| Electrical distribution system | 18 km |
| Electrical sub-stations | 33 |
| Raw water distribution system and storage system | 20 km |
| Potable water treatment plants | 2 |
| Sewage collection system, screening plant and submarine outfall | 1 |
| Accommodation camps capable of accommodating people in three classes of room together with village support facilities (Fig. 5) | 3 |
| Number of beds in camps | 8400 |
| Area canteens | 5 |
| Fire stations | 2 |
| Cottage hospital | 1 |
| Bus terminus | 2 |
| Paved roads | 25 km |
| Stormwater drainage system | 20 km |
| Rock suitable for crushing into aggregates | 8 Mt |
| Helicopter pad | 3 |

- its reliance on large numbers of imported workers
- there was a multitude of contractors working for many developers.

The AA chose to manage these factors positively and not merely allow contractors to fend for themselves. This was done by the AA constructing facilities and providing services which collectively were known as 'construction support' and the AA in effect became a management contractor who provided primary services to all contractors, developers and other third parties working on the project site. These services were provided by the AA when it was better placed to do so than contractors for reasons of economy of scale, safety and timeliness of availability. Construction support facilities that were provided are shown in Table 3.

The AA provided the following services free of charge to contractors which resulted in lower tender prices for construction contracts

- a chartered ferry service which delivered up to 12 000 workers to the site by boat daily (Fig. 6)

- a project site security system and personnel for identifying and screening all personnel arriving at the site
- a docking and marine cargo handling management system
- fire fighting and emergency rescue team available 24 hours a day, 365 days a year
- advice on safety, environment and industrial relations matters
- cleaning and maintenance of common areas.

The AA awarded a number of construction support licences (Appendix 1) following an open competitive tendering process. The successful licensees were able to sell goods and services at controlled prices. Generally, construction contractors were not compelled to use these services but could set up their own facilities (but not sell to others) if they chose to do so. Thus healthy competitive markets were established on site. The following licences were awarded

- a rock crushing licence to sell 8 Mt of crushed rock product
- two ready-mixed concrete batching plants

- two workforce accommodation camp operation and catering licences
- a primary health care service staffed with doctors and registered nurses
- two fuel suppliers
- a materials testing facility
- a waste disposal service
- a land transportation service
- printing services.

In addition, construction support provided assistance to the AA's own project staff thus relieving them of this burden so as to concentrate on their core mission of construction of the permanent works. Such support included the following

- operating and maintaining the AA's marine fleet of 14 vessels
- operating and maintaining the AA's fleet of 200 land-based vehicles
- maintaining the AA's portfolio of 100 site buildings
- operating and maintaining 200 radios and mobile telephones
- providing an escort service and logistics support to up to 2000 visitors who visited the construction site weekly.

## Construction contract packages

The framework for the contracting strategy was based on the *New Airport Master Plan*. There were four groups of developers which were responsible for the various construction contracts which comprised the overall airport project (Appendix 1)

- the AA
- the Hong Kong government
- the aviation franchisees
- the commercial partners.

The largest developer on the project was the AA. It had to provide a passenger terminal building, one runway together with associated taxiways and aprons, a road expressway, a railway corridor, a primary and secondary road distribution system, utilities, storm water systems, waste and grey water systems, communications systems, sea-cooling water systems and a ground transportation centre. The budget of design and construction of the AA's direct works was established in February 1994 at HK$ 44.447 billion and was never changed.

The Hong Kong government was responsible for the design and construction of the

- air traffic control complex
- government flying service facility
- air mail centre
- regional police headquarters
- landside fire station.

The government works department responsible

for the design and construction contracts was the Architectural Services Department (ASD). The Authority was responsible for management of co-ordination issues and interfaces and the provision of construction support services to ASD's contractors. Liaison with ASD was via the AA's representative and his delegates.

Franchisees were responsible for the design and construction of various aviation-related facilities which were essential to the operation of the airport and in which private enterprise was willing to make an investment. These facilities comprised

- two air cargo handling facilities
- three aircraft caterers
- one fuel delivery, storage and hydrant distributor
- a base maintenance operator
- three line maintenance operators.

Again, the AA's construction management role was primarily one of co-ordination and interface management. This was quite substantial as each site was an island within an island which had to form a cohesive whole. Each franchise agreement included a construction schedule which was administered by an AA representative who had powers to instruct variations, value such variations and award extensions of time to the franchisee. The franchisee entered into construction contracts with its own contractors.

Other business partners provided airport-related services; many of these were not essential to be in place before the opening of the new airport but were vital for its commercial success. Such enterprises included an hotel, airline headquarters, a freight forwarding centre and over 120 retail tenancies in the terminal building. These agreements also contained construction schedules and details of designated AA representatives who had similar powers to those in the franchise agreements.

The contracting strategy was continually adapted as funds were released to the PAA under the step-by-step financing arrangement which continued until both the government's financial support agreement was in place and the AA itself was properly established in December 1995. This meant that contracts were funded and awarded on a just-in-time basis in order to preserve the date for airport opening. This stop–go scenario, where contracts were awarded on a stand alone basis, was both more complex and more challenging to the AA than if the whole project could have been financed and planned as a total entity. This can be illustrated by the terminal building which was designed, and tender packages prepared, on the basis of four major components

- the structural and builder works
- the building services

*Fig. 7. Eastern vehicular tunnel being constructed beneath the southern runway (December 1995)*

- the fit-out works
- the communication systems.

In the event the terminal building was contracted quite differently with

- a foundations contract
- a superstructure contract
- the building services contract
- over 20 direct or nominated sub-contract fit-out packages
- nine direct systems contracts.

## Initial contracts

There were five contracts which were awarded early as it was envisioned that these needed to be brought to an advanced stage at the earliest opportunity, either because they had a long period for offshore development and manufacture or their site works had a significant impact on other works. The first contracts to be awarded following the SPC were the automated people mover and the baggage handling systems contracts. Both systems had to be developed to meet the unique demands of Chek Lap Kok and both

required a long offshore manufacturing phase. It was also important to award three civil works contracts early. The airfield tunnels were needed to take aircraft servicing vehicles under the southern runway, and it was necessary to award this contract in advance before the design of the runways and taxiways was complete (Fig. 7). Similarly, it was necessary to award an early contract for the construction of 25 km of storm water culverts. These were obviously in the deepest excavations on the site and had to be installed first. The third contract was to build a primary electrical substation for the electricity supply company—permanent electricity distribution was needed early for testing and commissioning the electrical equipment.

The construction packaging strategy was developed based on three main factors

- engineering discipline
- system or structural completeness
- geographical location.

The AA's general conditions of contract have a project manager (who is an AA employee) to

Fig. 8. *Completed passenger terminal building foundation and start of super-structure columns (September 1995)*

Fig. 9. *Roof module fabrication yard showing paint shop in centre and assembly jigs to the right (April 1996)*

administer each contract impartially. The duties and powers of the project manager are loosely based on those of the engineer under the Institution of Civil Engineers *Conditions of Contract*. The individual contracts were allocated to one of three project managers depending on the predominant engineering discipline of civil works, electrical and mechanical works or building works. Contracts were also sensibly packaged so that wherever possible a whole system would be the complete responsibility of a single contractor. Finally, the whole of the site was divided into four major geographical areas and a major contract was awarded for each of these, namely

- the terminal building
- the airfield works
- landside infrastructure works
- the ground transportation centre works.

Each of these geographical areas was under the control of one of the AA's construction managers, who in turn reported to a project manager. The construction manager was responsible for all co-ordination issues arising from all contractors with-

in the geographical area notwithstanding the contractor was in contract with the AA itself, the government or a business partner. Each geographical area had in the order of about 20 main contractors.

## Terminal building

The contract to build the foundations of the terminal building was awarded in June 1994. The works comprises 460 bored piles, 450 driven H-piles, 2600 tension anchors, a 280 m long multi-cell tunnel 36 m wide by 10 m high, 50 000 m² of ground slab and associated cast-in services (Fig. 8). The terminal building is located on the shore-line of the original island of Chek Lap Kok and extremely variable ground conditions were encountered. This resulted in variations to the design to the foundations as the works progressed. The original duration of the contract was set at 11 months but extensions of time were awarded and the actual construction period extended to 19 months.

In January 1995, the contract was awarded for the construction of the remainder of the passenger terminal building (PTB). At HK$ 10·1 billion it was the highest value contract awarded in the whole of the Airport Core Programme; however, it should be noted that almost 50% of this total was for prime-cost sums for sub-contracts to be awarded later once funding was available. The contractor was a joint venture comprising companies from the UK, the People's Republic of China, Japan and Hong Kong. The contract also included options for a scheme to widen the concourses, which increased the gross floor area from 490 000 m² to 516 000 m², and these were exercised within 28 days of the award of contract.

There was a relatively long mobilization period while the joint venture procured special table formwork, awarded sub-contracts for concrete works and obtained visas for imported workers. The first permanent works concrete was placed in May 1995 and peak production rates of 10 000 m³/week was achieved in December 1995. Ultimately, a total of 240 000 m³ of concrete was placed by the time concreting operations were completed in December 1996. By civil engineering standards this is not an exceptionally large amount of concrete but the total number of pours was in the order of 10 000 and gives some idea of the complexity of the building, and good planning was required to achieve these production rates. All concrete was provided using on-site batching plants and using Chek Lap Kok aggregates. The quality of the concrete was of a high standard and the great majority was fair-faced concrete with a painted surface coating.

One of the more technically challenging parts of the contract was the fabrication of the steel frame for the roof of the PTB. The roof structure was designed to be formed from orthogrid space frames forming barrel-vault modules designed to span 36 m from column to column. The bottom flange of each beam is exposed in its final condition and this architectural requirement imposed engineering constraints on the size of the frame members. It meant that the width and depth of every beam had to be the same and varying loads could only be catered for by varying the flange thickness. This meant that the quality of fabrication and welding had to be of a high standard. Steel for the roof was fabricated in England and Singapore prior to receiving a first coat of paint before being shipped to Hong Kong, where it was assembled. The contractor set up five jigs on site, the largest one being capable of holding a 36 m x 54 m long module with the rise on the barrel vault being 6 m (Fig. 9). The smooth-flowing alignment of the geometry of the roof demanded good dimensional tolerance control. The average

cycle time for the main assembly of a module was 23 days followed by six days for painting. A completed module was lifted from the jig by a 1000 t crawler crane and placed in an adjacent paint shop where it received two coats of white epoxy paint. The crane lifted it from the shop and placed it on two harnessed pairs of multi-wheeled trans-

*Fig. 11. (below and bottom). Roof module being transferred into position using launching beams traversing to the centre of the passenger terminal building (July 1996)*

Fig. 12. Roof of passenger terminal completed and weather-tight (June 1997)

porters, which took it to store where it was fitted out with purlins, roof sheeting, walkways and skylights.

In total 136 modules were assembled, each weighing anything between 40 and 140 t. The first module was lifted into position on 31 December 1995 using another 1000 t crawler crane (Fig. 10). A beam launcher supported by temporary towers was used to move modules laterally into the centre of the building where they could not be placed directly using the large cranes (Fig. 11). Individual members were then welded *in situ* to form a stitch between adjacent roof modules. The final module was lifted into position on 23 December 1996.

The roof covering comprised layers of metal decking, cementitious boarding to provide a sound insulation layer, thermal insulation, a vapour barrier and two layers of PVC waterproof membrane (Fig.12). The perimeter of the building was then glazed at departure level and a proprietary panel cladding system was fixed at arrivals and apron levels.

There are 38 frontal gates which comprise a

Fig. 13. Roof cantilevering 36 m out from the passenger terminal building to cover the departures forecourt

*Fig. 14. Check-in area (December 1997)*

bifurcated fixed link to separate arriving and departing passengers, each connected to pairs of apron drive airbridges. The fixed links were fabricated and erected *in situ* whereas the apron drives were manufactured offshore and brought complete to Hong Kong.

The building services contract was awarded on 30 January 1995 on the same day as the passenger terminal building contract. At HK$ 1·88 billion it was reputed to be one of the largest single building services contracts ever awarded. The works comprised four major disciplines of electrical, ventilation, fire and hydraulic services. These two major contracts were awarded on the same day because of the close dependency of one upon the other, both as to the need for the works to progress in parallel and for each contractor to assist the project manager to co-ordinate the design and construction of the works.

The tendering of nominated fit-out sub-contracts was delayed because funding was not available until the AA was established in December 1995. This resulted in the AA not being able to nominate the first packages until May 1996, at which time the contractor objected to the nominations collectively because of the lateness and their incompatibility with his own contract programme. This impasse was broken by the AA negotiating terms for the withdrawal of the notice of objection. These included a commercial settlement for all claims submitted up to July 1996 and agreement to work to a new programme to completion which would ensure an April 1998 airport opening date. A supplemental agreement was signed by the AA and the contractor in September 1996 which settled these matters. It was necessary to negotiate concurrently a separate agreement with the building services contractor to ensure both would be working to the same programme to completion.

Ultimately there were 22 nominated sub-contracts, two domestic sub-contracts and eight direct contracts for what were all originally included as prime-cost sums in the passenger terminal building contract. These, together with the

*Fig. 15. Dynamic compaction roller*

12 other direct contractors working for the AA in the PTB, gave the project manager's team an enormous task to carry out his duty to co-ordinate the design and construction activities of all contractors. For example, the nominated sub-contractor for the internal metal walling in public areas had to be given all the interface requirements by the project manager who gathered the information from all other contracts before the design of the panelling design could be finalized and be released for offshore manufacture. Some idea of the magnitude of this problem can be imagined when the total length of wall cladding to be installed under this package was over 40 km. The fit-out works were substantially complete by February 1998 (Figs 13 and 14).

## Airfield works

The contract for the airfield works was awarded in the sum of HK$ 2·4 billion on April 1995. The major works comprised the construction of the southern runway (3800 m long by 60 m wide), three parallel taxiways, two cross-field taxiways, the passenger aircraft aprons and the cargo aprons, giving a total paved area of 370 ha. The runway had to be complete within 17 months to allow instrument landing system calibration flights to start in November 1996 to match the programme for commissioning the overall air traffic control systems for the new airport.

The site preparation of the runway area had been left unprofiled and in flat platforms compacted to 95% standard Procter dry density. The first operation was to profile the earthworks by relocating 2 Mm³ of material and improving the density to 95% and 98% modified Procter dry density to depths of 2 m and 300 mm, respectively. The

earth was re-handled by using a mixed fleet of self propelled scrapers and articulated trucks loaded by back-hoes. Compaction of the marine sand was achieved by using up to 40 passes of a 20 t high-energy impact roller towed by four-wheel drive tractors (Fig. 15). The final compaction was carried out by watering and four passes of a towed 15 t vibrating roller. All formations were proven by the passage of a pneumatic-tyred 200 t proof roller, which simulated the maximum aircraft wheel loadings.

Runway pavement construction comprised a 400 mm thick layer of crushed aggregate base course (CABC). The specified minimum 10% fines values of 180 kN for this material could not be guaranteed from the aggregates arising from Chek Lap Kok, and the contractor had to import 3 Mt of aggregates from China to produce the CABC and other materials requiring this higher quality aggregate. The contractor set up a marine aggregate unloading facility to handle materials brought to the site by barge. From there the aggregates were blended to the right proportions using one of three pug mills, each having the capacity to process 200 t/h of material. The pug mills loaded product into the back of articulated dump-trucks which delivered the CABC to pavers which spread the material in specified thickness which were compacted using vibrating rollers.

The contractor set up a hot mix asphalt plant on site which had a capacity to produce 200 t/h, which was delivered to three paving machines. The total quantity of asphalt placed was 480 000 t and the maximum production recorded in a seven-day period was 12 000 t.

The third form of paving employed was concrete block paviors which were used where con-

Fig. 16.  *Southern runway viewed from the west (August 1997)*

Fig. 17. *Road and rail approaches to the ground transportation centre (December 1997)*

crete stands were required in areas of high residual settlements on reclamation. The blocks were manufactured in China to a strict tolerance so that the gaps between blocks were 2 to 4 mm to ensure good interlocking properties. The total quantity of blocks placed was 400 000 m² and the maximum production achieved in a seven-day period was 8000 m². Outputs were very significantly lower than this in times of rain and also where the areas were small, irregular in shape or had inserts such as manholes.

The runway contractor was also responsible for the fuel hydrant and distribution system works. This comprised 18 km of 600 mm dia. pipework with 13 mm wall thickness. Pipes were manufactured in 10 m lengths in the USA and sprayed internally with epoxy coating and wrapped with protective tape externally in Malaysia. The pipes were welded together at the side of the trench and then lowered into position following X-ray inspection of every weld. The whole system was charged with aviation fuel and tested to a pressure of up to 28 bar.

The main contractor was also responsible for the airfield ground lighting, the high mast light-

Fig. 18. Southern
commercial area
(December 1997)

ing and the grassing works. The ground lighting contractor had to co-ordinate closely the installation of the base cans, which house the lights, into the sub-base works as this proceeded. The asphalt was then placed before coring at the location of the cans and the light fitting installed. There were 7000 lights in the contract which were wired in series on a constant current with voltage up to 6000 V.

The first aeroplane to land on the new runway touched down on 20 February 1997. This was followed by a daily test plan whereby a Federal Aviation Administration aircraft made a series of landings, flights and takeoffs to test the air traffic control systems installed and operated by the Civil Aviation Department (Fig. 16).

### Ground transportation centre

The passenger terminal building and the airfield works are used to process and transport people who are travelling by air. There is an equal challenge to process and transport people on the ground; at Chek Lap Kok this is done at a ground transportation centre immediately adjacent to the passenger terminal. This allows people to arrive

and depart by high-speed train, by private car, by taxi, by tour coach or by bus. All these works were awarded to a single contractor on 1 January 1996 following the completion of a separate contract for the piling works.

The design and construction of the railway works on Chek Lap Kok were entrusted to the AA by the Mass Transit Railway Corporation (MTRC), which was responsible for the design, construction and operation of the whole of the new airport express railway. The railway station itself was of reinforced concrete and built on three floor levels with a concrete roof covered by aluminium cladding (Fig. 17). Passengers are enclosed in an air-conditioned environment by a curved glazed atrium and are able to walk across bridges and ramps to and from the passenger terminal building protected from the weather. The main contractor was also responsible for the building services works and the fit out works carried out by two nominated sub-contractors.

Other railway works constructed by the AA included a traction sub-station, the track formation at grade, the approach and departures viaducts and a cleaning platform at the end of the lines.

Fig. 19. Airport substantially completed (March 1998)

The track-related works were subsequently installed by the MTRC system-wide contractors.

The works also included the provision of facilities for motor vehicular transport. Vehicles will arrive at Chek Lap Kok by means of a six-lane expressway elevated to departures level by an approach ramp which is seven spans long, 20 m wide and 20 m above ground level. The lower deck is used as a transfer area for pedestrians to circulate. Below this is a service road at grade and below the service road lies the utilities corridor which provides all the services necessary to support the PTB. Construction of this part of the works was troublesome because of the multi-layer nature of the work which had to be carried out in a long and narrow congested works area with limited access.

The vehicles are brought down to ground level from kerb level by way of another post-tensioned concrete viaduct. Vehicles can then rejoin the expressway at grade, go to a multi-storey car park, a coach park or taxi stand or join the local at grade road distribution system. Arriving passengers find all their land transportation at facilities provided at grade which were constructed by the main contractor for this package of work.

## Infrastructure works

An international airport needs extensive infrastructure support facilities, notably a road and utility distribution system on the scale of a small town. A large contract to provide infrastructure

*Table 4. Infrastructure support facilities*

| Facility | Quantity |
| --- | --- |
| Ground level local roads | 25 km |
| Expressway | 5 km |
| Bridges | 5 |
| Communication ducts and associated pits | 101 km |
| Stormwater drains | 96 km |
| Manholes | 1400 |
| Ductile iron water pipes | 56 km |
| Sea water pipes | 24 km |
| Sewerage main | 17 km |
| Buildings including services and finishes | 40 |
| Security fencing | 9 km |
| Soft landscaping (excluding airside grass) | 135 km |

facilities to the south and eastern parts of the island was let in September 1995 (Fig. 18).

This contract was not technically difficult but the logistics and co-ordination management required provided the main challenges. Key quantities are given in Table 4.

In addition to this package the AA managed a number of other stand-alone construction contracts which had to be awarded individually for programming or logistical reasons. These were for the following facilities

- 33 kV primary electrical sub-station containing transformers and switch gear
- sea-water pumping station to house four 900 mm dia. pumps
- sea-water pumps (22 ranging from 70 l/s to 680 l/s)
- 33 kV electricity supply system for 81 MVA capacity
- 25 km of high voltage and 17 km of low voltage distribution
- 13 km of gas distribution system
- 1600 street lights and 193 km of associated cable works
- one grey-water treatment plant of 4900 m³/day capacity.

## Communications systems

A modern international airport requires an up-to-date integrated communications system. This comprises two major parts: that which is inside the passenger terminal building and that which cover the rest of the airport site. The overall design resulted in a seamless link between the two parts to produce a single integrated communication system. This design was packaged and let in ten contracts for the detailed design, manufacturing, installation, testing and commissioning. The packages and some of the items provided are (see Appendix 1)

- flight information display system
- public address system
- telephone system
- trunk mobile radio systems
- closed-circuit television system
- access detection and control system
- building management system and SCADA
- fibre-optic and copper cabling
- systems integration
- traffic surveillance and control system.

Basically, the system comprised a network linking four primary nodes, ten secondary nodes and fifty user nodes. A node in simple terms is a computer room which has been constructed by other contractors and fitted out with the standard of finish and facilities normally found in any computer room. The primary and secondary nodes were equipped with the airport operational database (AODB) computers which provided the nerve centres. The nodes are connected together by a local area network which comprises fibre optic and copper cables laid in the ducts and other cable containment provided by other contractors. This provided the backbone links which the other systems contractors could hook on to by connection into the cable termination points.

Each system was tested and commissioned on a standalone basis before they were all integrated and tested together. To assist this a special building was constructed and fitted out on site to carry out as much testing and integration as possible using simulation. This allowed much of the computer software to be debugged in advance of the hardware being transferred and installed in the works and helped shorten the period during which the systems works were on the critical path.

## Concluding remarks

The Hong Kong construction industry has an excellent record in overcoming enormous challenges in order to complete major projects on time and within budget. Nowhere is this better illustrated than the achievements seen on the airport project and on the Airport Core Programme as a whole. The challenges were made even greater when it was announced in late 1996 that construction of the second runway and an extension to the passenger terminal building should start before opening of the airport. Due to Hong Kong's continued economic growth, it is planned that the second runway should be operational before the end of 1998, to be followed shortly thereafter by opening of the north-west concourse extension to the passenger terminal building.

## References

1. LAM B. C. L. Management and procurement of the Hong Kong Airport Core Programme. *Proceedings of the Institution of Civil Engineers, Civil Engineering, Hong Kong International Airport Part 1: airport*, **126**, 1998, 5–14.
2. PLANT G. and OAKERVEE D. Hong Kong International Airport—civil engineering design. *Proceedings of the Institution of Civil Engineers, Civil Engineering, Hong Kong International Airport Part 1: airport*, **126**, 1998, 15–34.
3. *New Airport Master Plan Final Report*, Greiner Maunsell, 1991.
4. AYSON I. J. Survey aspects in the construction of Hong Kong's new airport of Chek Lap Kok, *Technical Proceedings of the 5th South East Asian and 36th Australian Surveyors Congress*, Singapore, 1995.

*Appendix 1: List of all contractors and nominated sub-contractors*

| Contract no. | | Contractor | Date of award |
| --- | --- | --- | --- |
| C023 | Kiosks and desks | Nederlandse Kunstof Industrie BV | Jan. 1998 |
| C031 | Fit-out of authority's offices in passenger terminal building | Collections Interiors Ltd | Oct. 1997 |
| C032 | Additional glazed screens in passenger terminal building | Seele GmbH and Co. KG | Dec. 1997 |
| C072 | Public area seating | Wilkhahn Wilkening and Hahne GmbH and Co. | Mar. 1997 |
| C075 | Broadloom carpet to public areas | Brintons Ltd | Mar. 1997 |
| C255 | Site investigation and laboratory testing | Intrusion Prepakt (Far East) Ltd | Sept. 1993 |
| C256 | Marine site investigation and laboratory testing | Lam Geotechnics Ltd | Oct. 1993 |
| C201 | Site preparation contract | Airport Platform Contractors JV, Comprising: Nishimatsu Construction Co. Ltd,Costain Civil Engineering Ltd, Morrison Knudsen Corporation, Jan De Nul NV, Bellast Nedam BV, China Harbour Engineering Co. | Nov. 1992 |
| C301 | Passenger terminal foundations | Gammon Construction Ltd–Nishimatsu Construction Co. Ltd JV | May 1994 |
| C302 | Passenger terminal building | The BCJ Joint Venture, Comprising: Amec Construction plc, Balfour Beatty Construction International Ltd, China State Construction Engineering Cooperation, Maeda Corporation, Kumagai Gumi (HK) Ltd | Jan. 1995 |
| | NSC365—lifts | Ryoden Lift and Escalator Co. Ltd | Feb. 1995 |
| | NSC366—escalators | Constructions Industrielles De La Méditerranée SA | Feb. 1995 |
| | NSC367—walkways | Constructions Industrielles De La Méditerranée SA | Mar. 1995 |
| | NSC370—aircraft loading bridges | PT Bukaka Teknik Utama–Ramp JV | Mar. 1995 |
| | NSC371—pre-conditioned air | Air-A-Plane Corporation | Mar. 1995 |
| | NSC422—ground transportation centre substructure | Leighton Contractors (Asia) Ltd | Apr. 1995 |
| | NSC066—glazed balustrades to public areas | MBM (HK) Ltd | June 1996 |
| | NSC067A—binnacles | European Manufacturing Joint Venture Ltd | July 1996 |
| | NSC067B—internal wall cladding systems, east of gridline II | Permasteelisa SpA (HK) | July 1996 |
| | NSC067C—internal wall cladding systems, west of gridline II | Permasteelisa SpA (HK) | Aug. 1996 |
| | NSC068—glazed screens and structural glazing | Seele (Hong Kong) Ltd | June 1996 |
| | NSC069—hard flooring to public areas | Grant Ameristone Ltd | July 1996 |
| | NSC070—metal suspended ceilings to public areas | G&H Montage (Hong Kong Project) Ltd | July 1996 |
| | NSC074—painting and decorating to public areas | SKK (Hong Kong) Ltd | Sept. 1996 |
| | NSC077—signage and graphics | Cevasa Imagen SA | Sept. 1996 |
| | NSC079—manufacture and supply of door furniture | Allgood Continental Ltd | June 1996 |
| | NSC080—landlord areas | YEK/Essman/KR Wong/Yiu Wing JV | July 1996 |
| | NSC083—public toilets | Yearfull Interior Contracting Co. Ltd | July 1996 |
| | NSC085—communications and control rooms | Vaford Contracting Co. Ltd | June 1996 |
| | NSC086—apm maintenance building | Chatwin Engineering Ltd | June 1996 |
| | NSC087—fixed link bridges | Shun Shing and Penta Ocean | July 1996 |
| | NSC088—check-in islands/reclaim units | Nederlandse Kunstof Industrie BV | Aug. 1996 |
| | NSC089—government areas | Vaford Contracting Co. Ltd | Sept. 1996 |
| C303 | North-west concourse piling | Franki Contractors Ltd | Nov. 1996 |
| C304 | North-west concourse and apron works | Zen Pacific–Shui On Joint Venture | Oct. 1997 |
| C320 | Passenger terminal building—building services | The AEH Joint Venture, Comprising: Aster Associate Termpoipianti SpA, Ellis Mechanical Services Ltd, Hsin Chong Aster Building Services Ltd | Jan. 1995 |
| C321 | North-west concourse—building services | Hsin Chong Aster Building Services Ltd; Aster Associate Termoimpianti SpA; Ellis Mechanical Services Ltd | Jan. 1998 |
| C350 | Automated people mover | New Hong Kong Airport People Mover System Joint Venture, Comprising: Sumitomo Corporation, Mitsubishi Heavy Industries Ltd, Kawasaki Heavy Industries Ltd, Nissho Iwai Corporation, Kobe Steel Ltd | Feb. 1994 |
| C360 | Baggage handling system | Siemens-Swire-Vanderlande | Feb. 1994 |
| C361 | Baggage trolley mechanical recirculation system | Associated Engineers Ltd | Nov. 1997 |
| C362 | Terminal building baggage trolley | Wanzl Metallwarenfabrik GmbH | Sep. 1997 |
| C365A | North-west concourse lifts | Ryoden Lift and Escalator Co. Ltd | Feb. 1998 |
| C366A | North-west concourse escalators | Constructions Industrielles De La Méditerranée SA | Feb. 1998 |
| C367A | North-west concourse walkways | Constructions Industrielles De La Méditerranée SA | Feb. 1998 |
| C370A | North-west concourse aircraft loading bridges | PT Bukaka Teknik Utama and Ramp International Incorporated | Feb. 1998 |
| C372 | Fixed ground power | Siemens Ltd | Apr. 1995 |
| C373 | Aircraft parking aids | Safegate International AB | May 1997 |
| C375 | Passenger and hand baggage security screening system | EG&G Astrophysics Research Corporation | Apr. 1997 |
| C376A | Baggage security screening system | Vivid Technologies Inc. | Dec. 1996 |
| C376B | Baggage security screening system | EG&G Astrophysics Research Corporation | Dec. 1996 |
| C377 | Level 3 baggage security screening systems | Invision Technologies Inc. | June 1997 |
| C378 | Level 3 baggage security screening trace detection equipment | Barringer Instruments Ltd | Sept. 1997 |
| C381 | Flight information display system | G.E.C. (Hong Kong) Ltd | June 1995 |
| C382 | Public address system | Hepburn Systems Ltd | June 1995 |
| C383 | Telephone system | Siemens Ltd | June 1995 |
| C384 | Trunked mobile radio | Marubeni Corporation and Hitachi Denshi Ltd | June 1995 |
| C387 | BMS and SCADA system | Control Systems International Inc. and Allen-Bradley (HK) Ltd | June 1995 |
| C388 | Voice and data cabling | International Computers Ltd | June 1995 |
| C395 | Closed circuit television (CCTV) system | Guardforce Ltd | June 1995 |
| C396 | Access control and detection | Guardforce Ltd | June 1995 |
| C399 | Master systems integration | Hughes Asia Pacific (Hong Kong) Ltd | May 1995 |
| C401 | Airfield works | Airfield Works Joint Venture, Comprising: Downer and Co. Ltd, Paul Y Construction Co. Ltd | Apr. 1995 |
| | NSC560—airfield ground lighting | Thorn Hong Kong | May 1995 |
| | NSC565—apron high mast | Au Chow Electrical Co. Ltd | Apr. 1995 |
| | NSC841—aviation fuel hydrant system | Wah-Chang Engineering Corp. Pte Ltd | June 1995 |
| | NSC470a—grassing | Gregori International SA | July 1996 |
| C402 | Northern runway works | ACG Joint Venture, Comprising: China Fujian Corporation for International Techno-Economic Cooperation, Heilit + Woerner Bau-AG. | Apr. 1997 |
| C410 | Landside infrastructure | Nishimatsu–Costain–China Harbour JV | Sept. 1995 |
| | NSC470b—grassing | Gregori International SA | July 1996 |
| | NSC472b—landscape planting and irrigationsystem, landside | Yee Sun Garden Ltd | Sept. 1996 |
| C420 | Ground transportation centre | Nishimatsu Construction Co. Ltd | Dec. 1995 |
| | NSC423—ground transportation centre, building services | UDL Kenworth Engineering Ltd | Jan. 1996 |
| | NSC425—ground transportation centre, escalators | CNIM Hong Kong Ltd | Jan. 1996 |
| | NSC426—ground transportation centre, walkways | CNIM Hong Kong Ltd | Jan. 1996 |
| | NSC427—ground transportation centre, fit-out, signage | | |

| | | | |
|---|---|---|---|
| | and graphics, FF and E | Vaford Contracting Co. Ltd | Aug. 1996 |
| | NSC470C—grassing | Yee Sun Garden Ltd | July 1996 |
| | NSC472C—landscape planting and irrigation system, GTC | Yee Sun Garden Ltd | Sept. 1996 |
| C430 | Airfield tunnels | Downer–Paul Y–McAlpine Joint Venture | Nov. 1994 |
| C451 | Airport maintenance facilities | Chun Wo Construction and Engineering Co. Ltd and China Civil Engineering Construction Corporation JV | Aug. 1996 |
| C452 | Sea rescue facilities | China State Construction Engineering Corporation | Jan. 1997 |
| C453 | Airport ancillary buildings | Wah Seng General Contractors Ltd | July 1996 |
| C454 | Airfield fire stations and control centre | Downer–Paul Y–McAlpine Joint Venture | July 1996 |
| C455 | Aircraft recovery equipment store | Dickson–China Harbour JV | Nov. 1996 |
| C456 | Jet blast screens | R&M Warme-, Schallschutz und Industrieservice Niederlassung/Branch Office München | Mar. 1996 |
| C471 | Grass trials | Gregori International SA | Feb. 1995 |
| C501 | Stormwater drainage box culverts | Hsin Chong Construction Co. Ltd | Nov. 1994 |
| C520A | Sea-water pumping equipment (north) | Young's Engineering Co. Ltd | May 1995 |
| C520B | Sea-water pumping equipment (south) | Young's Engineering Co. Ltd | July 1995 |
| C525 | Waste-water treatment plant | Sociedade de Construcciones Soares Da Costa SA; and WABAG Water Engineering Ltd | May 1996 |
| C531 | Primary substation A | Gold Banner Construction and Development Ltd | Sept. 1994 |
| C532 | Electrical distribution system | Balfour Beatty Ltd | Apr. 1995 |
| C534 | Emergency power equipment | The China Engineers Ltd | Apr. 1995 |
| C535 | Sea-water pumping house (north) | Nishimatsu–CHEC Joint Venture | May 1995 |
| C536 | Traffic control and surveillance system | G.E.C. (Hong Kong) Ltd | Jan. 1997 |
| C537 | Road lighting | Electricity Advisory Services Ltd | Sep. 1996 |
| C561 | Airfield ground lighting—northern runway | Thorn Lighting CLK Ltd | May 1997 |
| C902A | Initial workforce accommodation, employer's office and contractors' transit office | Shun Shing Construction and Engineering Company Ltd | May 1994 |
| C902B | Workforce accommodation and offices phase II (base contract) | Shun Shing Construction and Engineering Company Ltd | Nov. 1994 |
| C902C | Workforce accommodation and offices phase III (base contract) | Shun Shing Construction and Engineering Company Ltd | Nov. 1994 |
| C904 | Raw water submarine pipeline | Leighton–Lam JV | June 1994 |
| C906 | Temporary utilities, roadworks and bridges | China Fujian–Downer–McAlpine JV | June 1994 |
| C907 | Temporary ferry pier and berthing structures | UDL Kenworth Engineering Ltd | May 1994 |
| C940 | Rock crushing facility | Nishimatsu Construction Co. Ltd | Oct. 1994 |

**List of construction support licences**

| | | | |
|---|---|---|---|
| L01 | Operation of rock crushing facility | Nishimatsu Construction Co. Ltd | July 1995 |
| L02 | Concrete batching facility (south) | Costain Building and Civil Engineering Ltd | Dec. 1994 |
| L02A | Concrete batching facility (north) | Ready Mixed Concrete (HK) Ltd | Dec. 1994 |
| L03 | Servicing and cleaning of office and site staff accommodation, | Gardner Merchant Kelvin Hong Kong Ltd | Feb. 1996 |
| L04A | Temporary fuel facility | Mobil Oil Hong Kong Ltd | Aug. 1995 |
| L04C | Temporary fuel facility | Caltex Oil Hong Kong Ltd | Aug. 1995 |
| L05 | Waste disposal | The World Cleaning Services Co. Ltd | Aug. 1995 |
| L08 | Materials testing facility | MateriaLab Ltd | Apr. 1995 |
| L11 | Ferry services | The Hongkong & Yaumati Ferry Co. Ltd | Oct. 1994 |
| L12 | Medical services | Airport Medical Services Ltd | June 1995 |
| L15 | Operation of the northern workforce accommodation, canteens and associated facilities at CLK | Gardner Merchant Kelvin Hong Kong Ltd | Nov. 1994 |
| L15A | Operation of the southern workforce accommodation, canteens and associated facilities at CLK | Gardner Merchant Kelvin Hong Kong Ltd | Apr. 1995 |
| L17 | Temporary transportation services | CLK Bus Co. Ltd | Sept. 1995 |
| L18 | Printing services | UDO Holdings PLC/LDO Ltd Nevada | Feb. 1996 |
| L31 | Electricity supply | China Light and Power Ltd | TBA |
| L32 | Telecommunications services | Hong Kong Telephone Co. Ltd | Dec. 1995 |

**List of government contracts**

| | | | |
|---|---|---|---|
| 711 | Government flying services | Architectural Services Department | Oct. 1995 |
| 712 | Air mail centre | Architectural Services Department | Jan. 1996 |
| 713 | Airport police station | Architectural Services Department | Oct. 1995 |
| 714 | Sub-divisional fire station | Architectural Services Department | Oct. 1995 |
| 730 | Air traffic control complex and tower (ASD Contract SSC 307) | Architectural Services Department | Oct. 1994 |

**List of all commercial franchisees**

| | | | |
|---|---|---|---|
| 530 | Permanent electricity supply | China Light and Power Co. Ltd | TBA |
| 540 | Town gas distribution system | Hong Kong and China Gas Co. Ltd | Feb. 1996 |
| 551 | Telephone exchange building | Hongkong Telecom Ltd; New World Telephone Ltd; Hutchison Communications Ltd | Feb. 1997 |
| 806 | Aircraft base maintenance facility | Hong Kong Aircraft Engineering Co. Ltd | Oct. 1996 |
| 811 | Air cargo facility No. 1 | Hong Kong Air Cargo Company Ltd (HACTL) | Dec. 1995 |
| 812 | Air cargo facility No. 2 | Asia Airfreight Terminal Co. Ltd (AAT) | Jan. 1996 |
| 814 | Ground support equipment maintenance facility No. 1 | Dah Chong Hong & Dragonair Airport GSE Service Ltd (DAS) | Feb. 1997 |
| 815 | Ground support equipment maintenance facility No. 2 | Ground Support Engineering Ltd | Mar. 1997 |
| 816 | Business aviation centre | Hong Kong Business Aviation Centre Ltd | July 1997 |
| 821 | Aircraft catering facility No. 1 | Cathay Pacific Catering Services (HK) Ltd (CPCS) | May 1996 |
| 822 | Aircraft catering facility No. 2 | LSG Lufthansa Service HK (KSG) | Mar. 1996 |
| 823 | Aircraft catering facility No. 3 | Gate Gourmet Hong Kong Ltd | Feb. 1996 |
| 830 | Nitrogen storage facility | Hong Kong Oxygen & Acetylene Company Ltd | Aug. 1997 |
| 840 | Aviation fuel facility | Aviation Fuel Supply Company (AFSC) | Dec. 1995 |
| 845 | Landside petrol filling stations (south) | Shell Hong Kong Ltd | May 1997 |
| 846 | Landside petrol filling stations (north-east) | Shell Hong Kong Ltd | May 1997 |
| 847 | Airside filling stations 1 to 3 | Mobil Oil Hong Kong Ltd | Feb. 1997 |
| 848 | Into-plane fuelling No. 1 | AFSC Refuelling Ltd | Mar. 1997 |
| 849 | Into-plane fuelling No. 2 | AMR Airline Services Fuelling (HK) Ltd | Jan. 1997 |
| 850 | Freight forwarding centre | Airport Freight Forwarding Centre Co. Ltd | Feb. 1996 |
| 856 | Cathay Pacific headquarters and crew hotel | Airline Property Ltd | May 1996 |
| 857 | CNAC/Dragonair headquarters | Hong Kong Aviation and Support Services Ltd and Profit Venture Development Ltd | Aug. 1997 |
| 858 | Cathay Pacific stores building | Airline Stores Property Ltd | May 1997 |
| 859 | Cathay Pacific flight training centre | Airline Training Property Ltd | May 1997 |
| 870 | Airport hotel | Bauhinia Hotels Ltd | Mar. 1996 |
| 871 | Multi-storey car park | Bauhinia Hotels Ltd | Mar. 1996 |

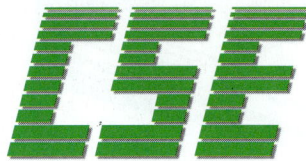

# Hong Kong International Airport

# A civil engineering success story

## Edited by: G W Plant, C S Covil, R A Hughes – Airport Authority Hong Kong

In 1990 the Hong Kong Government launched the Airport Core Programme to provide a new international airport, together with nine associated projects, at a total cost of over twenty billion US dollars, making it one of the world's largest infrastructure developments. The completion of the new Hong Kong International Airport in only seven years and within budget is testament to Hong Kong's remarkable record of delivering major construction projects. This book presents a record of the design, construction and performance of the airport platform.

*Site preparation for the new Hong Kong International Airport* describes how two islands and a vast area of sea were converted into a 1,248 hectare airport platform site in just two and a half years. The sensitivity of the local environment required that all environmental aspects of the work were completely planned and controlled.

Written and edited by leading practitioners who played key roles in the successful completion of this huge project, this book is not merely a record of the works but highlights the many complex issues that the construction team faced, right from the design through to the completion of the airport platform.

*Site preparation for the new Hong Kong International Airport* clearly presents both the theory and practice for the dredging and land reclamation works and provides valuable lessons in site investigation, geotechnical instrumentation, surveying and settlement prediction and analysis. This book will quickly establish itself as an invaluable reference work for all engineers and academics who are involved in civil infrastructure, maritime works, geotechnics, quarrying/mining, dredging and the environment — in fact anyone who is interested in learning about one of the major civil engineering projects of this century.

Site preparation for the new Hong Kong International Airport
Design, construction and performance of the airport platform

Edited by G W Plant C S Covil and R A Hughes